Law and Economics of Vertical Integration and Control

Law and Economics
of Vertical Integration and Control

ROGER D. BLAIR
Department of Economics
University of Florida
Gainesville, Florida

DAVID L. KASERMAN
Department of Economics
University of Tennessee
Knoxville, Tennessee

1983

ACADEMIC PRESS
A Subsidiary of Harcourt Brace Jovanovich, Publishers
New York   London
Paris   San Diego   San Francisco   São Paulo   Sydney   Tokyo   Toronto

ACADEMIC PRESS, INC.
111 Fifth Avenue, New York, New York 10003

United Kingdom Edition published by
ACADEMIC PRESS, INC. (LONDON) LTD.
24/28 Oval Road, London NW1 7DX

Library of Congress Cataloging in Publication Data

Blair, Roger D.
  Law and economics of vertical integration and
control.

  Includes index.
  1. Trusts, Industrial--United States. 2. Industry
and state--United States. I. Kaserman, David L.
II. Title.
HD2795.B548 1983   338.8'042   83-8793
ISBN 0-12-103480-1

PRINTED IN THE UNITED STATES OF AMERICA

83 84 85 86   9 8 7 6 5 4 3 2 1

This book is dedicated
to Chau and Lois

Contents

Part II Legal Analysis of Vertical Control   137

Part III   Public Policy Analysis   187

10 Conclusion

Acknowledgments

In writing this monograph, we have accumulated many debts that can be repaid only partially by acknowledging them. First of all, we owe an enormous intellectual debt to the scholars who have led the way. Almost all of them are cited in our bibliography. There are a few, however, who had a great influence on our thinking: Robert Bork, Meyer Burstein, Ronald Coase, George Hay, Richard Schmalensee, F. R. Warren-Boulton, and Oliver Williamson. We have also benefitted from discussions with our colleagues, Sid Carroll, Tom Cooper, and Dave Qualls. We thank all of them most warmly.

Second, we have drawn freely upon some of our previous publications. We wish to thank the publishers of the *American Economic Review, The Antitrust Bulletin, Economics Letters,* and the *Southern Economic Journal* for permission to use portions of previously published material. In addition, we want to thank the publishers of *Economica,* the *Journal of Industrial Economics,* the *Journal of Law and Economics,* and the *Journal of Political Economy* for permission to publish excerpts of previously published materials by others.

We have received generous financial support from the Public Policy Research Center at the University of Florida. In addition, Kaserman received faculty research grants from both the University and the College of Business at the University of Tennessee. In addition, the graphics were prepared under the able supervision of Marjorie Niblack at the University of Florida. Almost all the typing was cheerfully and expertly done by Frances Kiernan Rhinesmith.

We enjoyed the editorial support and production assistance provided by our association with Academic Press.

Finally, we owe a great deal to the support, encouragement, and motivation provided by our families, who do not care one whit about vertical integration—contractual or otherwise.

Introduction

Background

The subject of vertical integration and vertical contractual restraints has been an intellectual battleground on which debate between lawyers and economists has continued for a long time. Like the fields of the Mekong Delta, this particular piece of terrain has seemed destined to host an almost continuous struggle between opposing armies throughout time. Innumerable skirmishes and several major campaigns have taken place concerning the appropriate public policy response to the various mechanisms by which firms at one stage of production influence the behavior of firms at a vertically related stage.

Evidence of the longevity of this struggle may be found throughout the literature on the subject. Approximately three decades ago, Robert Bork (1954, p. 201) wrote, "A comparison of the law and the economics of vertical integration makes it clear that the two bear little resemblance." Almost 20 years later, George Hay (1973, p. 188) indicated that the battle is continuing: "There is probably more disagreement among lawyers and economists on the subject of vertical integration, especially vertical mergers, than in any other area of antitrust." And, after another 3 years of conflict, John McGee and Lowell Bassett (1976, p. 18) were still unable to detect any movement toward a truce: "Whether because of mis-emphasis or not, academic and legal controversy about vertical integration has been going on for a long time, and still continues. Even so, it is not obvious whether and how much progress has so far been made."

The latest news from the vertical control front, however, appears guardedly optimistic. Some major breakthroughs in the economic theory of vertical integration and control have occurred over the last decade which have improved our understanding of the causes for and consequences of these practices. One can argue that these new discoveries have encouraged some economists to take a slightly more conciliatory view of the legal profession's historical opposition to vertical integration. At the same time, some recent judicial and administrative decisions may have signalled a considerable reduction in the level of hostility toward vertical integration and control exhibited by the courts and the antitrust enforcement agencies. Thus, while it is still too soon to be certain, it appears that some convergence has begun to occur between the views held in these opposing camps.

The extent to which economists' thinking on the subject has changed can be seen in the recent shift in the theory of forward integration by an intermediate-product monopolist. For a long time, it was generally believed that an upstream monopolist (possibly exempt from antitrust attack by virtue of being a labor union or by the existence of one or more patents) would have no incentive to integrate into the production of downstream products (see, e.g., Spengler, 1950, and Machlup and Taber, 1960). The only exceptions to this rule occurred when (1) significant transaction costs associated with the use of the market mechanism made it cheaper to transfer the intermediate product to downstream producers internally (Coase, 1937); or (2) the downstream industry was also monopolized (Spengler, 1950).[1] In either of these situations, it can be shown that the effect of vertical integration is not only to increase the profits of the upstream monopolist, but also to expand output and reduce price in the final product market. Consequently, in both cases, overall social welfare is unambiguously improved by bringing the separate stages of production within the control of the intermediate-product monopolist, that is, by allowing (or possibly even encouraging) vertical integration. On the basis of these results, the economics profession generally argued that the appropriate antitrust policy regarding vertical integration by an input monopolist was one of conciliation at best or neutrality at worst.

In 1971, however, John Vernon and Daniel Graham demonstrated that, even in the absence of transaction costs or downstream monopoly, an input monopolist could increase profits by vertically integrating into the final-product market if the production function for the final-good industry exhibits variable input proportions.[2] This article touched off a series of papers showing that (1) a basic profit incentive exists for intermediate-product monopolists to integrate forward into competitive downstream

---

[1] An additional potential incentive that has been subject to debate for many years involves a possible effect of vertical integration on barriers to entry. We ignore this incentive here but shall return to it in a subsequent chapter.

[2] This result actually appeared quite a bit earlier in Burstein (1960). For some inexplicable reason, it went unnoticed.

industries; (2) the profit incentive to integrate persists until the intermediate-good monopolist has successfully monopolized the final-good industry; and (3) given the existence of monopoly power at the upstream stage, the social welfare effects of a successful monopolization of the downstream industry through vertical integration are a priori indeterminate.[3] Thus, the verdict of innocence that the economics profession had returned earlier regarding vertical integration by an intermediate-product monopolist has been overturned by subsequent literature. The potential for negative welfare effects has modified the antitrust policy recommendations of most economists regarding the appropriate treatment of vertical integration by an input monopolist from a policy of per se legality to a rule of reason approach under which each case is considered separately.

On a related front, there has been an increasing awareness among economists of the isomorphic nature of vertical control by ownership and by contract. Publications, for example, by Blair and Kaserman (1978a, 1980), have demonstrated the economic equivalence between vertical ownership integration by an intermediate-product monopolist and various contractual arrangements observed in interindustrial relationships. These contractual alternatives to ownership integration include: (1) an input-tying arrangement, (2) an output royalty, (3) a sales revenue royalty, and (4) a lump-sum entry fee. These generic contractual arrangements may be seen to correspond to a host of specific contract terms incorporated in patent licensing agreements, labor union contracts, and franchising systems. Recognition of their economic equivalence has led several economists to recommend a more equal treatment of these business practices under the antitrust laws. Posner (1981) has gone a step further and has called for a policy of per se legality of all vertical control arrangements.

This basic trend in the economics profession toward a more wary yet consistent attitude concerning vertical control appears to be mirrored by concurrent developments in the legal treatment of vertical restraints. Although a more enlightened legal reasoning is only tentative at this point, it now appears that the judiciary is beginning to both recognize the commonality of alternative vertical restraints and admit the possibility of other than pernicious effects. This emerging trend may be seen in the history of the legal treatment of tying arrangements.

Prior to the passage of the Clayton Act in 1914, tying cases had to be brought under the Sherman Act. These efforts were largely unsuccessful. Section 3 of the Clayton Act, however, provided the foundation for an

---

[3] Although Vernon and Graham (1971) unquestionably renewed our interest in vertical control, these papers by Hay (1973), Schmalensee (1973), and Warren-Boulton (1974) spawned a large addition to the economics literature.

increasingly hostile attitude toward tying arrangements. One of the best-known judicial assessments of tying is contained in Justice Frankfurter's opinion in the *Standard Stations* case: "Tying arrangements serve hardly any purpose beyond the suppression of competition." [See *Standard Oil Co. of California (Standard Stations)* v. *United States* 337 U.S. 293, 305–306 (1949).]

In 1953, the Supreme Court laid down the different standards of proof of illegality under the Sherman and Clayton Acts. In *Times-Picayune,* the Court held that the Sherman Act condemns a tying arrangement whenever (1) "sufficient economic power" is shown in the tying good and (2) a "not insubstantial" amount of commerce in the tied goods is affected. [See *Times-Picayune Publishing Co.* v. *United States,* 345 U.S. 594 (1953).] In contrast, the Clayton Act is offended when *either* of these conditions is satisfied. Subsequent decisions so attentuated the requirements for proving either condition that there appeared to be practically no distinction between the two statutes.

This deterioration in the standards of proof culminated in the first *Fortner* decision. [See *Fortner Enterprises, Inc.* v. *United States Steel Corp.,* 394 U.S. 495 (1969).] Here, it was held that if an "appreciable number of buyers" submits to a tying arrangement, then one can infer the existence of sufficient economic power. At that point, it seemed that tying was virtually a per se offense. But this impression has changed with the second Fortner decision, [*United States Steel Corp.* v. *Fortner Enterprises, Inc.,* 429 U.S. 610 (1977)] in which the Supreme Court made it far more difficult for the plaintiff to sustain the burden of proof necessary for condemning a tie-in sale as per se illegal.[4] Milton Handler (1977) was moved to hail this decision as a reinstatement of the rule of reason in tying cases.

In the same year, the *Sylvania* decision was announced. [See *Continental T.V. Inc.* v. *GTE Sylvania,* 433 U.S. 36 (1977).] This decision followed a decade of severe criticism of the Supreme Court's decision in *Schwinn*. In *Schwinn,* the Court had decided that nonprice vertical restraints on customer selection were per se illegal. There followed a decade of evasion by the lower courts and criticism by the academic community by both lawyers and economists. [See *United States* v. *Arnold, Schwinn & Co.,* 388 U.S. 365 (1967).] In *Sylvania,* the defendant had imposed a location clause upon its authorized dealers. Thus, Sylvania had employed a nonprice vertical restraint in its distribution system. The Court acknowledged that the *Sylvania* facts made the *Schwinn* rule applicable but concluded that the *Schwinn* rule had been a mistake. In Justice White's concurring

---

[4] For somewhat contrasting views on this point, see Jones (1978) and Baker (1980).

opinion there is an unmistakable recognition of the considerable similarity between nonprice vertical restraints and vertical price fixing.

Thus, with economists admitting the possibility of negative welfare effects, with lawyers admitting the possibility of positive welfare effects, and with both sides becoming increasingly aware of the isomorphic nature of alternative vertical control mechanisms, the emerging legal and economic attitudes display some tendency to converge. Additional evidence supporting this convergence hypothesis may be found in the revised *Merger Guidelines* issued by the Antitrust Division of the Department of Justice. [See Department of Justice Release, Merger Guidelines, June 4, 1982; reprinted 1069 ATRR pp. 5-1–5-16 (June 17, 1982).] These guidelines indicate a significantly more liberal attitude toward vertical mergers than was exhibited in the preceding set of guidelines. [See Department of Justice Release, Merger Guidelines, May 30, 1968; reprinted, 360 ATRR pp. X-1 et seg. (June 4, 1968).] Moreover, there is a notable absence of any reference to the infamous foreclosure doctrine, which has long been the primary weapon used in the legal profession's attacks against vertical integration and control.

Whether this convergence of views will continue until an armistice is reached is highly uncertain. The fundamental policy prescriptions advocated by economists have not been altered dramatically, and legal resolutions are notoriously transitory.[5] But, if convergence should continue, it may well represent one of the most significant advances made in the realm of public policy in many decades. In this monograph, we shall examine the emerging concensus and evaluate our current public policy posture.

Purpose of Present Study

The literature on the law and economics of vertical integration and control is vast and scattered. In this monograph, we attempt to present a comprehensive and coherent survey of this literature. We try to make corrections where necessary and expand the analysis where appropriate. As we develop the economic case for vertical integration in response to varied circumstances, we also examine contractual alternatives. In each instance, we try to determine the extent to which the contractual alternatives are economically equivalent to ownership integration.

[5] Bork (1978a, p. 171), for example, has remarked that the *Sylvania* decision "is either the most important and promising antitrust decision of the past two or three decades or merely the latest inconclusive episode in the Court's continuing travail in the wilderness of the law of vertical restraints."

In the second part of the monograph, we turn our attention to public policy. Although vertical integration and control can take many forms, these alternative forms provide, in many situations, economically equivalent results. In these situations, one should expect all forms to be treated similarly by the antitrust laws. This, however, is not the case. Many vertical control mechanisms receive hostile treatment while others are treated leniently. Ownership integration, vertical price fixing, and tying arrangements are per se illegal or nearly so. In contrast, exclusive dealing, customer and territorial allocations, and requirements contracts are subject to a rule-of-reason standard. Finally, output and sales royalties as well as lump-sum franchise fees are presumptively legal. Most of these vertical control arrangements are economically equivalent to each other, and, as such, should be treated in the same way. This part of our study will develop and critique the judicial attitude toward these various business practices. Along the way, we provide guidance for the appropriate judicial policies concerning vertical restraints.

Standard for Analysis

The fundamental standard of benchmark that we use in evaluating the various vertical control mechanisms is consumer welfare. If a business practice causes output to expand, then there is good reason to believe that consumers are better off as a result. Such practices ought to be encouraged rather than discouraged. In cases where output is unchanged by a business practice, the appropriate public policy is benign neglect. Even when output is reduced by a particular business practice, we cannot assert categorically that it should be proscribed. Output reductions, however, raise a red flag and suggest that a close look should be taken.

Considerable scholarly research supports our assertion that consumer welfare ought to be our benchmark for evaluating vertical control mechanisms. In spite of some unfortunate Supreme Court language to the contrary,[6] it appears that consumer welfare was of paramount concern to the drafters of the original antitrust legislation. Thorelli's (1954) impressive history of the Sherman Act provides ample support for this position. His analysis is based upon extensive quotes from Senator Sherman's defense

---

[6] In *Brown Shoe Co.* v. *United States,* 370 U.S. 294, 344 (1962), the Supreme Court claimed that Congress was willing to sacrifice efficiency in order to preserve "small, locally owned businesses." This, of course, would be inconsistent with consumer welfare since higher costs would cause prices to rise. Turner (1965), however, has pointed out that no credible support exists for the Court's contention in *Brown Shoe.*

of his bill. These quotes display Sherman's deep concern for the consumer. Thorelli's review of the Sherman Act's legislative history led him to the conclusion that Congress advocated more competition. Moreover, he felt that Congress perceived that the ultimate beneficiary of the competitive process was the consumer.

Letwin (1956), an astute economic historian, noted that one of the various trust abuses that led to passage of the Sherman Act was consumer victimization caused by high prices. Finally, Bork has treated this issue extensively (1966) and has analyzed the purpose of the antitrust laws from two perspectives (1978b). The first perspective is the declared legislative intent, which, according to Bork's analysis of the legislative history, clearly indicates that Congress's exclusive purpose was the promotion of consumer welfare. Furthermore, the courts were not to balance consumer welfare against social values. The second perspective is the legislative intent that can be inferred from a structural analysis of the statute. This, too, reveals an overriding concern for consumer welfare. Thus, the consensus of Thorelli, Letwin, and Bork is compelling: the antitrust laws originally purported to protect and promote consumer welfare. We shall adopt this purpose in analyzing vertical integration and control.[7]

---

[7] For an argument that noneconomic objectives of antitrust are important, see Pitofsky (1978).

PART I

Economics of Vertical Control

Transaction Costs

In agreement with Morris Adelman (1949, p. 27), we shall say that a firm engages in vertical integration when that firm "transmits a good or service which could, without major adaptation, be sold in the market." As this definition makes clear, the distinguishing feature of vertical integration is the replacement of a market exchange by an internal (within the firm) transfer. In the former case, resource allocation is governed at the aggregate level by the market supply of and demand for the intermediate product and at the disaggregate level by the bilateral negotiation and contracting process that occurs between individual buyers and sellers. In the latter case, resource allocation is governed by the unilateral administrative decisions of the managers of the firm and the bureaucratic or hierarchical processes through which these decisions are implemented.

The existence of these two alternative mechanisms for coordinating the allocation of productive resources was first elaborated by Ronald Coase (1937, p. 333):

> Outside the firm, price movements direct production, which is co-ordinated through a series of exchange transactions on the market. Within a firm, these market transactions are eliminated and in place of the complicated market structure with exchange transactions is substituted the entrepreneur-co-ordinator, who directs production. It is clear that these are alternative methods of co-ordinating production.

Identifying these two fundamental systems for directing the allocation of resources, Coase established an economic definition of the firm: "the distinguishing mark of the firm is the supersession of the price mechanism." Thus, without any vertical integration at all, firms would not exist.

Intermediate products, capital, and labor would be joined through a system of market contracts between resource owners. Production decisions would evolve from multilateral negotiations among these owners rather than the centralized process of managerial decision making that exists within a firm. Moreover, the size of the individual firm is viewed by Coase as being governed by the number of intermediate product markets that are, by its existence, internalized.

Assuming, then, that firms are organized for the purpose of earning profits, a (if not the) primary reason for replacing a market exchange with an internal transfer must be that, for that particular input, the latter mechanism is less expensive than the former: "The main reason why it is profitable to establish a firm would seem to be that there is a cost of using the price mechanism" (Coase, 1937, p. 336). The costs of using the price mechanism generally have come to be called transaction costs. Thus, transaction costs refer to any expenditure of resources associated with the use of the market in transferring a good service from one party to another.

As Schupack (1977, p. 4) has pointed out, Coase's analysis of transaction costs and vertical integration provides a good starting point for understanding the organizational decisions of the firm, but it falls somewhat short of a complete treatment in two important respects. First, Coase provides little detail concerning the underlying sources of transaction costs. This void hinders any predictive use of the analysis; that is, it would be difficult on the basis of Coase's work to predict which particular classes of transfers would be likely candidates to be organized internally versus across a market or how such candidacy might be expected to change over time. And second, Coase's treatment implicitly assumes competitive intermediate-product markets, so that market structure influences on and consequences of vertical integration simply do not arise. This, in turn, makes any analysis of the welfare effects of vertical integration incomplete if not impossible. The first of these shortcomings will be addressed in this chapter. The latter will be dealt with in the remainder of the book.

The organization of the rest of this chapter is as follows. First, we show how the presence of transaction costs can provide a profit incentive for vertical integration under competitive market conditions. Next, we indicate what the welfare effects of such integration are. Following that, we survey some of the more recent literature in order to provide additional detail concerning the origins of transactions costs. Then, we briefly describe the potential advantages of internal transfers. And finally, we discuss some contractual alternatives to vertical integration in the presence of significant transactions costs.

The Incentive to Integrate

To enhance our understanding of the role of transaction costs in influencing the decision of a firm to internalize a given transfer of an intermediate product, we shall find it useful to present our arguments graphically. In Figure 2.1 we have drawn the value of the marginal product curve VMP_x and the marginal cost curve MC_x for an individual buyer and an individual seller of the intermediate product x, respectively. Assuming competition in the purchase and sale of this intermediate product, these curves represent the demand curve of the buying firm (VMP_x) and the supply curve of the selling firm (MC_x).

Then, as Arrow (1969, p. 60) indicated, "In a price system, transaction costs drive a wedge between buyer's and seller's prices." Due to the expenditure of resources required to carry out the given transaction, the effective price paid by the buyer will exceed the effective price received by the seller. For simplicity, we shall assume that transaction costs are a constant amount per unit of the intermediate product transferred from the seller to the buyer.[1]

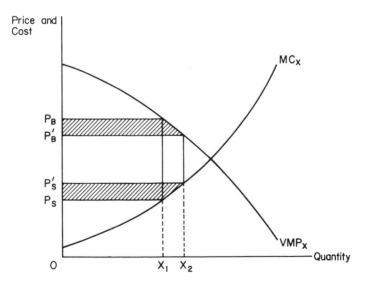

Figure 2.1 Profit incentive to integrate due to transaction cost savings.

[1] As Arrow (1969, p. 60) pointed out, due to the fixed cost nature of information acquisition expenditures, there is good reason to expect the per-unit transaction costs to decline with the size of the individual transaction, that is, with the number of units of the intermediate good traded. Incorporation of this aspect of transaction costs, however, would complicate but not alter the substance of our argument here.

Let this amount be equal to $P_B - P_S$ in the graph if the transaction is conducted across the market. Market exchange will, then, lead to an equilibrium transfer of X_1 units of the input, with the buyer paying P_B and the seller receiving P_S.[2] Profits to the buyer from engaging in this transaction are given by the area above P_B and below the VMP_x curve. Profits to the seller are given by the area below P_S and above the MC_x curve.

Obviously, the higher the transaction costs, the lower will be the market quantity traded. In the extreme, transaction costs can become so great that the product will not be traded at all. As Arrow (1969, p. 60) notes, this case corresponds to the usual notion of market failure: "market failure is the particular case where transaction costs are so high that the existence of the market is no longer worthwhile." Barring this extreme case, however, the market will continue to exist, but the quantity exchanged will be constrained by the magnitude of the transaction costs.

We shall assume that the costs of organizing the transfer of intermediate good x from the upstream producer to the downstream producer are reduced from $P_B - P_S$ to $P'_B - P'_S$ if the given transfer is internalized, that is, if the producing and purchasing agents are combined within a single firm. As many authors have pointed out, internal transfer costs may be less than market transaction costs in many situations, but they are not likely to be zero. This is because "although the human and transactional factors which impede exchanges between firms (across a market) manifest themselves somewhat differently within the firm, the same set of factors applies to both" (Williamson, 1974, pp. 1442–1443). Thus, we expect $P'_B - P'_S > 0$ to hold for internal transfers as well.

Given this assumed cost saving, the total increase in profits due to vertical integration of the buying and selling firms is given by the sum of the two shaded areas in the graph. Clearly, as long as $P_B - P_S > P'_B - P'_S$, this area will be positive. That is, the basic profit incentive to engage in vertical integration will exist whenever internal transfer costs are less than market transaction costs. Moreover, the strength of this incentive (i.e., the magnitude of the profit increase that will result from vertical integration) will vary directly with: (1) the size of the reduction in the costs of organizing the exchange [i.e., $(P_B - P_S) - (P'_B - P'_S)$]; and (2) the elasticity of both the VMP_x and the MC_x curves.

This last point does not appear to be of any real significance in the simple model presented above because the binary decision of whether to integrate or not seems to depend entirely upon a comparison of $P_B - P_S$

[2] Here, we have chosen to represent these costs as a wedge between the demand and supply curves. Alternatively, one could shift the MC_x curve upward or the VMP_x curve downward by the amount $P_B - P_S$. Either of these approaches would provide equivalent results.

with $P'_B - P'_S$. If the former exceeds the latter, the profitability of vertical integration is assured, and the firms involved will be expected to integrate regardless of the actual magnitude of the profit increase.[3] The apparent simplicity of this decision rule, however, stems from two assumptions that are implicit in the above model. First, it is assumed that the costs of internal transfer, $P'_B - P'_S$, are completely exogenous to the firm. And second, it is assumed that the firm makes the integration decision independently for each intermediate product market in which it operates. In fact, neither of these assumptions accurately portrays the complex organizational decision problem confronted by the firm.

To begin with, the actual costs of organizing a given exchange internally are expected to vary directly with the number of exchanges that the firm has already internalized. As Coase (1937, p. 340) points out "as a firm gets larger, there may be decreasing returns to the entrepreneur function, that is, the costs of organizing additional transactions within the firm may rise." Thus, while it may be reasonable, as a first approximation, to assume that market transaction costs, $P_B - P_S$, are exogenous to the firm, it is clearly unreasonable to treat internal transfer costs, $P'_B - P'_S$, as a given.

Once this interdependence between the cost of internal transfers of one intermediate product and the organizational choices made for other intermediate products is recognized, the complexity of the optimization problem faced by the firm in deciding what set of exchanges to internalize can begin to be appreciated. From a set of n intermediate products employed by the firm, it must select some subset that will minimize the total costs of exchange, recognizing that the per-unit cost of transferring each product internally depends upon how many (and, indeed, which) of the others are selected for internalization. In this setting, the elasticities of the relevant VMP and MC curves for each input can become quite important. In fact, it may be optimal for the firm to internalize an exchange for which the per-unit transfer cost saving is less than that of other exchanges that might alternatively be internalized. An example will help to clarify this point.

Suppose a firm employs two intermediate products, x and y. Let the per-unit market transaction costs associated with x equal $P_B^x - P_S^x$ and the per-unit market transaction costs associated with y equal $P_B^y - P_S^y$, where we assume that $P_B^x - P_S^x > P_B^y - P_S^y$. Let the per-unit internal transfer costs of x, denoted by $P_B^{x'} - P_S^{x'}$, equal zero if the y transfer is not

---

[3] The magnitude of the profit incentive to integrate (and, therefore, the elasticities of the VMP_x and MC_x curves) would be of crucial importance if the firm faced the risk of antitrust attack by vertically integrating. The decision to integrate, then, would depend upon a comparison of such profit and risk. If the profit increase were small and the threat of legal action were large, the firm might choose not to integrate despite the fact that $P_B - P_S > P'_B - P'_S$.

internalized but exceed the per-unit market transaction costs, $P_B^x - P_S^x$, if the y transfer is internalized. Finally, let the per-unit internal transfer costs of y, denoted by $P_B^{y'} - P_S^{y'}$, equal zero if the x transfer is not internalized but exceed the per-unit market transaction costs, $P_B^y - P_S^y$, if the x transfer is internalized. Summarizing our assumptions, we have

(1) $\quad P_B^x - P_S^x > P_B^y - P_S^y$

(2) $\quad P_B^{x'} - P_S^{x'} \begin{cases} = 0 \text{ if } y \text{ is not internalized,} \\ > P_B^x - P_S^x \text{ if } y \text{ is internalized,} \end{cases}$

(3) $\quad P_B^{y'} - P_S^{y'} \begin{cases} = 0 \text{ if } x \text{ is not internalized,} \\ > P_B^y - P_S^y \text{ if } x \text{ is internalized.} \end{cases}$

Given these assumptions, only one of the two intermediate product exchanges will be selected for internalization. This is because internalization of either input raises the internal transfer costs of the other input above that input's market transaction costs. The optimal input to select for internalization, however, cannot be determined solely from the three assumptions stated above. Additional information concerning the VMP and MC curves for both inputs must be obtained.

Suppose these curves appear as in Figure 2.2, where the elasticities of both curves are greater for the y input than for the x input. Then, in this case, it will be optimal for the firm to integrate across the y market and leave the x market nonintegrated. The increase in profits from integrating y, leaving x nonintegrated, is given by the shaded area in the right-hand graph, while the increase in profits from integrating x, leaving y nonintegrated, is given by the shaded area in the lefthand graph. Clearly, the former is greater than the latter, so the firm will integrate across the market that exhibits relatively less savings in transaction costs per unit.

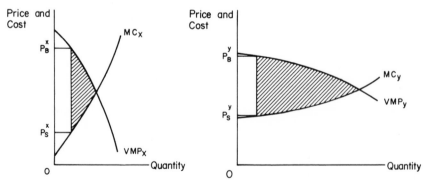

Figure 2.2 Comparative increase in profits from integrating the x versus the y input.

The purpose of this example is to show the inadequacy of the simple decision rule which states that a firm should always integrate across any market for which internal transfers are cheaper than market transactions. The effect of such integration upon the costs of internal transfers of other intermediate products must also be considered, and the fully simultaneous optimization problem can obviously become quite complex when more than two inputs are utilized. In this example, we found that the firm would integrate across the market for which the per-unit reduction in the costs of completing the exchange were smaller but the total savings were larger. This result, too, could easily be altered if we introduce differential impacts on the costs of internal transfers of the other inputs. In other words, it is conceivable that optimality would require integration across a market in which the total cost savings for that particular input were less if such integration has relatively less impact upon the internal transfer costs of other inputs.

The Welfare Effects

Despite the complexity of the preceding optimization problem, the firm that is successful in solving it will have selected its organizational structure so that the overall costs of transferring *all* intermediate products from the upstream to the downstream stages of production are at a minimum. To the extent that this solution involves the replacement of some market transactions with internal transfers, it is clear that vertical integration undertaken for this purpose improves social welfare. Returning to Figure 2.1, the vertical integration of input x not only increases the producer's profits but also expands employment of the integrated input from x_1 to x_2. Since such expansion is necessarily accompanied by an expansion of final output (since the marginal product of the input is positive), consumers will benefit as well.

The conclusion that vertical integration undertaken to reduce transaction costs is welfare enhancing must, of course, be tempered if significant market structure effects result from such integration. As Schupack (1977, p. 6) pointed out:

> A close analogy can be drawn between the economies of vertical integration and the more familiar scale economies of large volume production. In both cases social welfare is served by taking advantage of the production [or transaction] efficiencies involved as long as the resource misallocation costs incurred by possible increases in market power are not too great.

In later chapters, we will examine the possible relationships between market structure and vertical integration. For now, we merely indicate that such relationships can exist and that they may either strengthen or negate the conclusion that integration will improve welfare.

We now turn to a discussion of those factors that operate to determine the magnitude of market transaction costs.

Transaction Cost Determinants

Coase's (1937) original analysis does not entirely ignore the various sources of transaction costs. Search costs are mentioned: "the most obvious cost of 'organizing' production through the price mechanism is that of discovering what the relevant prices are" (p. 336). In addition, some of the problems associated with the use of long-term contracts are briefly described: "It may be desired to make a long-term contract for the supply of some article or service. . . . Now, owing to the difficulty of forecasting, the longer the period of the contract is for the supply of the commodity or service, the less possible, and indeed, the less desirable it is for the person purchasing to specify what the other contracting party is expected to do" (p. 337). But, the most extensive treatment of transaction cost determinants by far has been carried out since Coase's analysis by Oliver Williamson. His fundamental contribution is his elaboration of the underlying causes of transaction costs and his development of the significance of such costs not only to vertical integration theory but to other areas of antitrust and microeconomics as well.

Williamson (1974) identified two sets of factors whose elements interact to increase the costs of market exchange. Since market exchange generally involves the use of contracts of varying lengths and complexities, these factors can be seen as impediments to the negotiation and enforcement of contractual agreements between buyers and sellers of an intermediate product. The first set of factors, which Williamson referred to as "transactional factors," is concerned with the environmental characteristics of the relevant intermediate product market. Specifically, the degree of market uncertainty and the number of potential trading partners available to the firm are discussed. The second set of factors, which Williamson referred to as "human factors," pertains to two rather ubiquitous characteristics of human nature that impede negotiations between parties with conflicting interests. These characteristics are labeled "bounded rationality" and "opportunism." A brief discussion of each of these four basic factors follows.

First, the term "market uncertainty" may refer to the future price or quality of the intermediate product or to the availability of either supply or demand at any price/quality combination (i.e., rationing on either side of the market). The fact that some degree of uncertainty must be present for transaction costs to arise was recognized by Coase (1937, p. 338): "It seems improbable that a firm would emerge without the existence of uncertainty." Malmgren (1961, p. 401) also emphasized the role of uncertainty in increasing the costs of market exchange: "if events were predictable the price mechanism would render its signalling service at no cost."

In the following chapter, we shall review several models that focus explicitly upon the effects of various specific types of uncertainty on the incentive for vertical integration. Here, we direct our attention only to the effects of uncertainty on the costs of carrying out exchange across the market. In general, the greater the degree of uncertainty that exists in a market, the more lengthy and complex will be the contracts that are negotiated between buyers and sellers of the intermediate product. Such complexity is required to guard the trading parties against changes in the market that might alter the precontract incentives to perform in the agreed upon manner. Since nonperformance by either party reduces the profits of the other party, both will want to specify the details of the contractual requirements as completely as possible. This, in turn, increases the costs of both negotiating and enforcing the contract. It also reduces the flexibility of the overall operation to adapt to changing market conditions. For both of these reasons, market uncertainty increases transaction costs.

Second, decreases in the number of potential trading partners available to the firm are likely to increase the costs of market exchange. What Williamson refers to as "small numbers bargaining problems" arise when traders' options for transferring their business to alternative suppliers or buyers are limited. Moreover, as Williamson (1971 and 1974) pointed out, there are many transactions that might involve a large number of potential traders when the original contract is negotiated but a very small number of potential traders when this original contract is renewed. This reduction in the number of potential traders at the contract renewal may be due to (1) specific investments made by the winner of the original contract; or (2) the creation of firm-specific human capital as a result of carrying out the terms of the original contract. For either of these reasons, the original contractor may obtain a real cost advantage that effectively prohibits other firms from successfully bidding on later contracts.

Third, the term "bounded rationality" was defined by Williamson (1970, p. 956) as a "condition in which human agents are '*intendedly* rational, but only *limitedly* so.' " Thus, the term signifies a type of behavior that falls in the middle ground between irrational or random action and the superrational calculating behavior often attributed to "economic man."

Here, economic agents are assumed to pursue their goals in a logical and consistent manner while experiencing rather severe limits on their abilities to receive, store, and process the phenomenal amount of information that would be required to attain these goals precisely. Because of these limits, "uncertainty" may exist at the individual level even when all relevant data are theoretically available, that is, when no market uncertainty is present.

And fourth, "opportunism" implies a type of behavior in which individuals attempt to realize gains "through a lack of candor or honesty in transactions" (Williamson, 1973, p. 317). Thus, while the concept of bounded rationality implies that individuals will behave in a less calculating manner than that implied by standard microeconomic theory, the notion of opportunism suggests a more sophisticated decision process which includes the pursuit of "self-interest with guile" (Williamson, 1979, p. 957). In situations where deception can be expected to increase profits, honesty in trading is not likely to persist.

Having defined each of these four basic factors, Williamson (1971, 1973, 1974, 1979) went on to describe how they interact with each other to increase the costs of the bilateral negotiation and contracting process involved in market exchange. In addition, he demonstrates how internalization of that exchange can, in many circumstances, lead to a reduction in these costs. In essence, Williamson's argument is that (1) long-term contracting costs are increased by the combined effects of market uncertainty and bounded rationality, while (2) the costs of relying upon an equivalent series of short-term contracts are increased by the combined effects of small numbers bargaining and opportunism.

Firms on both sides of an intermediate-product market make investments in productive equipment with the expectation of being able to earn a return that is sufficient to amortize the investment and yield a surplus that at least covers the opportunity costs of the funds that are committed. Realization of this expected return generally depends upon some specific performance on the part of a firm or firms on the other side of the market over an extended period of time. Sellers of an intermediate product would not invest in the productive capacity required to manufacture this product if they did not expect to be able to sell their output to downstream firms at or above some given price. At the same time, the downstream producers that require this input would not invest in productive equipment if they did not expect to be able to purchase the intermediate product in sufficient quantities at or below some given price. Consequently, both buyers and sellers will be interested in taking steps to ensure that their expectations regarding the postinvestment behavior of the other party will be met. To a large extent (though not entirely) this is achieved through the bilateral negotiation of contracts that specify each party's behavior over the contract period.

Long-term contracts specify such behavior over an extended period of time. Obviously, this type of contract is relatively appealing when the investment that the contract is designed to protect is long lived. The presence of market uncertainty, however, makes it difficult to specify the desired behavior in advance because the optimal behavior depends upon the outcomes of future events that are not known at the time the contract must be negotiated. Consequently, if the contract locks the parties into a certain pattern of behavior that is optimal at the time the contract is designed, it is very likely that this same pattern of behavior will prove to be suboptimal for one or both of the parties in some future period. Rigidly specified contracts thereby impose costs in the form of lost flexibility to adapt behavior to unforeseen circumstances.

This problem could, theoretically, be addressed by making the specified behavior contingent upon the actual outcome that is manifested in the future periods. But, since the number of potential outcomes is generally quite large, a complete specification would result in a prohibitively complex contract. Given the universal attribute of bounded rationality on the part of all negotiating parties, it is doubtful that such a contract could be designed in many situations, let alone negotiated and enforced. The information storage and processing capabilities of the human agents involved are simply insufficient to the task in a world where the future is uncertain and the term of the contract is long. In somewhat less complex situations in which it is feasible to specify and negotiate such a contract completely, the costs of doing so may be quite high.

In addition to specifying and negotiating the long-term contract, the parties involved must also be concerned with postcontractual enforcement. This involves both detecting and punishing behavior that is not in compliance with the terms of the contract. The former problem may be far from trivial if the product involved is subject to subtle quality variations. And the latter problem must be of serious concern if the parties rely solely upon litigation to discourage postcontractual opportunistic behavior. Given the existence of uncertainty regarding future market conditions, there is likely to be a nonzero probability that at least one of the parties involved will find it optimal to renege on the contractual agreement entirely. Certainly, the contract itself cannot guard against violation of its own terms. Taking recourse to legal means of enforcing compliance can be both costly and uncertain and therefore cannot eliminate the risk of nonfulfillment of the contractual obligations.

In situations where contractual reneging is a serious concern (and where vertical integration is not an attractive alternative), the market mechanism generally incorporates some sort of economic, as opposed to legal, sanctions or safeguards that are designed to discourage or to minimize the adverse effects of noncompliance. [See Klein, Crawford, and Alchian (1978) Goldberg (1979) and Klein (1980) for analyses of market responses to the risk of contractual reneging.] Examples of this sort of market adaptation are fairly common. The opportunistic party may suffer a loss of future business through a damaged reputation. It may forfeit a lump-sum payment that was made at the contract negotiation stage or its rights to a future income stream from continuing the contractual arrangement. The party facing the threat of contractual reneging may invest in some sort of back-up facilities or maintain what would otherwise be an unnecessarily large inventory of the affected intermediate good. The use of these adaptive devices, however, further increases the costs of relying on the market mechanism for the transfer of the intermediate product.

Long-term contracts thus face the twin problems of providing flexibility (which is required because of uncertainty) while, at the same time, avoiding complexity (which is costly because of bounded rationality). Therefore, where market uncertainty is great, it is unlikely that long-term contracts will be extensively employed. In these situations, the firms involved may (and, in some cases, do) turn to the use of an equivalent series of short-term contracts. These have the advantage of providing needed flexibility at contract renewal intervals while avoiding the necessity of specifying in advance responses to all possible contingencies. In the words of Williamson (1971, 1979), a series of short-term contracts allows an "adaptive, sequential decision-making" process to unfold while, at the same time, permits the firm to "economize on bounded rationality."

Short-term contracts, however, may prove to be unattractive in situations in which what Williamson refers to as "small-numbers bargaining problems" arise. By this, Williamson simply means that the party initiating the contract is limited, for any of a variety of reasons, to a small set of potential contract partners. Prior to this party's making an investment in productive capacity, such limitation is not a serious impediment to the operation of the market mechanism. The decision to go forward with the investment can be conditional upon finding a trading partner willing to provide acceptable terms. Following such investment, however, the firm becomes vulnerable to opportunistic behavior in which the firm's contract partner alters the terms of the agreement at contract renewal. In short, the fixed investment made by one party generates a stream of quasi-rents that may be appropriated by the other party. [Klein, Crawford, and Alchian (1978) define quasi-rents as the value of an asset in its current use that is in excess of its salvage value.] Of course, were it not for the limitation on the number of potential trading partners, the firm would simply switch to an alternative supplier or purchaser when such opportunistic behavior arose. In the absence of alternative trading partners, however, the firm may have no choice but to surrender a portion or all of the quasi-rents generated by the prior investment. This is precisely the same problem as that described by Klein, Crawford, and Alchian (1978), except that, with short-term contracts, the contract expiration and subsequent renegotiation make more blatant contractual reneging unnecessary. Consequently, the problem of such opportunistic behavior is more likely to arise where a series of short-term contracts is employed.

Williamson's analysis of these four basic factors, then, predicts that the costs of using the market mechanism to coordinate the exchange of an intermediate product is likely to be high where market uncertainty is relatively great and where short-term contracts involve (or are likely to lead to) small-numbers bargaining. In these situations, the firms are apt to find that the costs of internal transfers are lower than the costs of market exchange.

Advantages and Limitations of Internal Transfers

Williamson's work on the ontology of transaction costs includes some analysis of the properties of internal transfers that, in many situations, make them less costly than market exchanges. A review of this work indicates four major properties that favor internal control over market forces.

First, where the intermediate product is produced by the same firm that employs it, a considerable degree of flexibility to adapt to changing market conditions is realized. Desired output and capacity changes can be readily communicated between members of the same firm so that adjustment lags may be reduced. Also, subtle styling or quality variations can be more readily accommodated. Plan consistency, or what Malmgren (1961) referred to as a "convergence of expectations," can be realized between the outputs of the two related stages of production. Consequently, the outcome-adaptive behavior that is sought through either contingent contracts or a series of short-term contracts is more readily available through an internalization of the relevant transfer.

Second, internal transfers are likely to reduce the opportunistic tendencies of the parties to the exchange. This results from the fact that "internal divisions do not have preemptive claims on profit streams" (Williamson, 1974, p. 1446). Instead, as members of the same firm, these parties have an incentive to maximize the joint profits of the overall operation. As a result, they will not be tempted to pursue a course of action that will benefit one party (or division) at the expense of the other. A good example of how internalization alters the incentive to behave opportunistically is provided by Williamson's (1974) discussion of the bilateral monopoly situation. Under this market structure, it is well known that the two monopolists will agree to exchange a quantity of the intermediate product that will maximize total industry profits. Otherwise, the two firms will not be on the contract curve, that is, one could be made better off without making the other worse off (Bowley, 1928). Once this quantity has been determined, however, the share of the maximized profit that accrues to each party will depend upon the price that is negotiated. In this situation, each party will have an incentive to expend resources on the price negotiation process up to the point at which the marginal cost of further negotiation equals the marginal revenue that such negotiation is expected to yield. This expenditure of resources on negotiation results in a reduction in the joint profits of the two firms involved. Vertical integration of these firms eliminates the incentive to bargain over the price of the intermediate good and, thereby, increases the total net profits that are available.

Next, although internalization reduces the incentive to engage in opportunistic behavior, it does not eliminate it altogether. The same lack of a preemptive claim on profits that reduces the incentive to increase the revenues of a given division of the firm at the expense of another division of that firm also reduces the incentive to hold down the costs of each division individually, because any increase in profits that results from cost reductions will be shared by all divisions. Consequently, opportunism survives within the firm but it takes on a somewhat different and, perhaps, more subtle form. Here, the third property of internal control discussed by Williamson operates to keep such opportunism within reasonable bounds. Specifically, the firm has available a much wider variety of options for controlling its internal affairs than it has for controlling the behavior of other firms with which it deals. In Williamson's words, "the firm possesses a comparatively efficient conflict resolution machinery" (1971, p. 114). The ability of a manager to command and reward desired behavior or to punish undesired behavior efficiently enables the firm to exercise much tighter reign over opportunistic inclinations that are manifested within the firm than those that arise in its dealings with other firms.

Fourth, a primary reason the firm is able to exercise greater flexibility and apply its superior reward and penalty mechanism in a more efficient manner is the existence of certain informational advantages that vertical integration confers. Information flows between related stages of production are likely to improve when these stages are combined within a single firm for several reasons. First, any reduction in the incentive to behave opportunistically reduces the threat of what Williamson (1971, p. 117) referred to as "strategic misrepresentation risk." Consequently, information that is passed from one stage to the other is more likely to provide an accurate representation of the true situation when these stages share common ownership. Second, common experiences by members of the same firm tend to facilitate communication of a more informal nature. Such communication improves the overall flow of information within the firm. And third, the firm has much greater access to the relevant performance data of its internal divisions than it has to the equivalent data of the firms with which it deals. For the former, the information may be obtained directly through internal audits; while, for the latter, inferences must be drawn from trading experiences.

The above properties tend to make the costs of internal transfer less than the costs of market exchange in many situations. As Coase (1937, p. 340) explained, however, this cost advantage may be expected to disappear as the firm's size increases: "as a firm gets larger, there may be decreasing returns to the entrepreneur function, that is, the costs of organizing additional transactions within the firm may rise." This explanation of the limitations of internal control, then, is founded upon the notion of managerial diseconomies of scale that stem from the bounded rationality of the firm's managers. A very similar explanation is offered by Williamson (1973, p. 323): "spans of control can be progressively extended only by sacrificing attention to detail. Neither transactional economies nor effective monitoring can be achieved if capacity limits are exceeded." Malmgren (1962, pp. 417–420) made comparable arguments. Thus, the prevailing view regarding transactional forces that may impede the process of internalization has remained remarkably constant over a considerable period of time.

Contractual Alternatives

The body of literature dealing with transaction costs generally treats all possible contractual arrangements under the broad heading of market exchange and proceeds to explain why these arrangements fail to provide incentive structures, information channels, or control apparatus that are equivalent to those realized under ownership integration. It is obvious, therefore, that no contractual arrangement exists that will duplicate exactly all of the transactional features exhibited by the vertically integrated firm.

At the same time, however, many contracts may be seen as market adaptations designed to address some specific transactional deficiency or set of deficiencies associated with pure spot market purchasing and selling. Long-term contracts are often designed to resolve supply or demand reliability problems. An equivalent series of short-term contracts might represent a compromise between reliability and flexibility to adapt the relationship to changing market conditions. Exclusive dealing and requirements contracts economize on search costs and at least partially address certain externalities that arise with pure spot market exchange of particular products. Finally, franchising agreements and what Goldberg (1979) classified more generally as "relational contracts" represent more complex market adaptations that are largely designed to alleviate simultaneous opportunistic inclinations on the part of both trading parties. (See also Caves and Murphy, 1976, Klein, Crawford, and Alchian, 1978, and Klein, 1980.) These various contractual arrangements are employed where *both* spot market trading and outright vertical integration are relatively costly methods for organizing the transfer. The use of these contracts, then, arises when transaction costs derive primarily from one or a few fairly specific sources and managerial diseconomies prohibit internalization of the given transfer.

This view suggests that the proximity of the transactional relationships that exist between the producer and the user of an intermediate product lies on a continuum, with spot market exchange and common ownership providing the extremes and the myriad contractual and noncontractual agreements falling in the middle.[4] The metric that varies as we move from the one end of this continuum to the other is the degree of control that one of the parties to the exchange exercises over the other. With pure spot market trading, no such control is exercised. Market forces (or the invisible hand) guide all resource allocation decisions. With vertical integration, on the other hand, the maximum feasible control is achieved. Owners of the integrated operation make all relevant production decisions.

This framework is valuable in analyzing the social welfare effects of vertical integration and control because it facilitates a logical separation of (1) the underlying incentives of firms at one stage to influence the production decisions of firms at another vertically related stage; and (2) the firm's selection of a specific control instrument to achieve such influence (ownership integration or contractual arrangements). A failure to make this separation leads one dangerously close to the conclusion that

---

[4] Other writers have recognized that the vertical relationships we observe in practice are not always easily categorized according to a binary market versus non-market taxonomy: "Many long-term contractual relationships (such as franchising) blur the line between the market and the firm" (Klein, Crawford, and Alchian, 1978, p. 326). Also, see Blois (1972).

all (or virtually all) vertical integration is in response to transaction costs, in which case one might also be led to conclude that all (or virtually all) vertical integration is welfare enhancing. While the latter may or may not be a reasonable conclusion, its support should come from a careful analysis of the welfare effects of the underlying motivation to exercise vertical control in the first place, rather than a secondary analysis of the firm's selection from among the available control instruments.

For example, we shall later find that an input monopolist that sells its output to a competitive downstream industry employing this input with others in variable proportions will have a profit incentive to influence the input mix of its customers. (See Burstein, 1960, Vernon and Graham, 1971, Schmalensee, 1973, and Hay, 1973.) This influence may be achieved through either vertical integration or a variety of contractual arrangements (see Blair and Kaserman 1978a, 1980, 1982a, 1982c, and Inaba, 1980), and the social welfare effects of the monopolist successfully exercising this influence are a priori indeterminate. (See Warren-Boulton, 1974, Mallela and Nahata 1980, and Westfield, 1981.) Now, we may presume that the input monopolist will opt for the vertical integration alternative only if the transaction costs associated with that alternative are less than those associated with the equivalent contractual arrangements. In this limited sense only has vertical integration occurred in response to transaction cost considerations. A welfare analysis that compares only the vertical integration versus contractual agreement alternatives will lead us to the erroneous conclusion that integration must improve welfare in this situation. The relevant comparison for welfare purposes here, however, is that between vertical control and no vertical control, which leads us to the correct conclusion that the welfare effects are indeterminate on theoretical grounds. Consequently, the logical separation of the firm's incentive to exercise vertical control from its selection from among the available control instruments is necessary if we are to assess the social welfare consequences of the observed behavior correctly. In the remainder of this book, we shall be concerned primarily with the incentive question.

3

Fixed Proportions and Contractual Alternatives

A striking characteristic of many vertical restraints is that they involve products that do not change physically as they move from the manufacturer through the distributor to the retail customer. For example, retail sellers of television sets, refrigerators, automobile tires, electric typewriters, newspapers, stereo equipment, and myriad other consumer goods simply perform a distribution function. These sellers do not alter the product physically, which is not to say that no useful function is performed. These sellers add important services to the wholesale product in converting it to a retail good. In this situation, there is a so-called fixed proportions relationship between one of the inputs and the output. In other words, for every television set sold to a retail customer there must be one television set sold by the manufacturer to the distributor. Under some circumstances, there is no need for vertical control, but other conditions make vertical control desirable to one or more of the parties involved in the manufacture-distribution chain.

In this chapter, we examine three situations that provide an incentive for vertical control under fixed proportions. These are (1) successive monopoly, (2) importance of product-specific services, and (3) entry barriers. In each of these cases, we discuss vertical ownership integration as well as a contractual alternative. We shall develop as a benchmark a simple model of derived demand where there is no incentive for vertical control. From there, we shall move to a consideration of the more complicated cases that call for vertical control.

No Incentive for Vertical Control

At the outset, we shall analyze the case in which the distribution industry is competitively organized. We shall suppose, however, that the manufacturer enjoys some horizontal market power due, say, to a patent. (The source of this market power is really not important, but we attribute it to a patent so that our attention is not deflected to the existence of horizontal market power at the manufacturing stage.) Accordingly, we shall assume that the manufacturer's market power is legal. The manufacturer sells his product to an extensive network of competitive retail distributors that, in turn, sells the product to the final consumers. Throughout this chapter, we shall assume that transactions are costless. As middlemen in the distribution scheme, the retail distributors demand the product from the manufacturer only to the extent that final consumers demand it from them. Consequently, we speak of the distributors' demand as being derived from the consumer demand. This derived demand is the demand function that dictates the optimal price and output decision of the manufacturer.

In Figure 3.1, the retail customers' demand for the final product is represented by D_R, which shows the usual negative relationship between

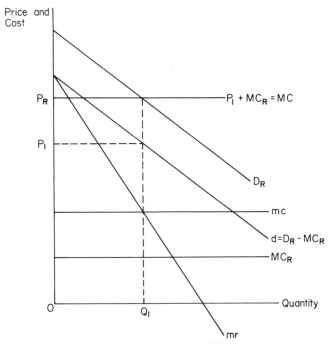

Figure 3.1   No incentive to integrate with competition at the downstream stage.

price and quantity. We assume that this demand function is nonstochastic. The per-unit (or average) cost of performing the retail function by the distributor is assumed to be constant with regard to the quantity sold. When this is the case, the marginal (or incremental) cost is also constant and equal to the average cost. Although the incremental retailing cost is assumed to be constant for expositional convenience, this poses no theoretical inconsistencies. Constant marginal costs are consistent with free entry and exit at the retail distribution level plus competitive input markets for the retail distribution industry as a whole. We assume that these distribution costs are not product-specific. In Figure 3.1, we represent the marginal cost of retailing by MC_R.

The demand for the manufacturer's product manifested by the retail distributors is derived from the retail consumers' demand for the product. The retail demand curve D_R shows the maximum price that consumers are willing to pay for the final good. Given the final demand D_R in Figure 3.1, the retailers cannot pay the manufacturer more than the price that consumers are willing to pay minus the costs of performing the retailing function. In other words, for any given quantity of final product, the maximum price that the retailers can afford is given by the height of the retail demand curve D_R at that quantity minus the marginal cost of retailing that quantity of output MC_R. Consequently, the derived demand confronting the manufacturer is d where $d = D_R - MC_R$. The marginal revenue function associated with d is shown as mr.

We assume that the manufacturer is interested in maximizing profits. In an effort to accomplish this objective, the manufacturer's optimal strategy is to select an output such that marginal production costs are equal to the marginal revenue associated with the derived demand. In Figure 3.1, the manufacturer's marginal production cost mc is assumed to be constant. Such constant marginal production costs can result from production in a single plant according to a linearly homogeneous production function combined with competitive input markets. Alternatively, constant marginal production costs can result from multiplant production and competitive input markets. The manufacturer's optimal (i.e., profit-maximizing) output is found where marginal revenue equals marginal cost. In Figure 3.1, this is shown as Q_1. This output will be sold at a price of P_1 to the competitively organized distributors. As a consequence, the manufacturer earns a monopoly profit of $(P_1 - mc)Q_1$.

Due to competition among the retail distributors, the price to the final customer will be equal to the price charged by the manufacturer plus the marginal cost of retailing. Since the vertical distance between the retail demand curve D_R and the derived demand d is precisely equal to the marginal cost of retailing MC_R, the final retail price shown in Figure 3.1

denoted by P_R is equal to P_1 plus MC_R. Note that the price P_R is a market clearing price for a quantity of Q_1. The competitive retailers earn a competitive rate of return since their price P_R is just equal to their costs: P_1 plus MC_R.

Under the conditions that we have specified (i.e., fixed proportions in production, competition at the distribution stage, no transactions costs, nonstochastic demand, and no product-specific services), all of the monopoly profit has been extracted by the manufacturer through the price and output decision regarding the intermediate product. That is, the final product price and quantity are equal to those that would result from a vertically integrated monopoly that controlled both manufacture and distribution. In other words, $P_R - mc - MC_R = P_1 - mc$. Since $P_1 = P_R - MC_R$, this result is immediately apparent. Consequently, there is no need for vertical control by the manufacturer. (See Spengler, 1950, and Machlup and Taber, 1960.)

Successive Monopoly

For many products, retail distribution is carried out by franchisees that have some local monopoly power. In some instances, the manufacturer assigns exclusive territories to the franchisees. The classic example of this involves the distribution of newspapers (see Blair and Kaserman, 1981). The newspaper publisher (manufacturer) has often assigned specific routes (exclusive territories) to newspaper carriers (franchisees) for home delivery service. In other cases, exclusivity is not guaranteed, but the manufacturer spaces the franchisees in such a way that each one can be a viable business entity. In other words, the cost structure of the franchisee's business indicates that excessive intrabrand competition among distributors would lead to failures. The classic example of this involves the distribution of automobiles. (See the excellent discussion by Pashigian, 1961.) A local market must be larger than some critical size or it will not support more than one, say, Buick dealer. This is because the minimum volume of sales necessary for financial well-being is fairly substantial. It is not in the interest of General Motors to have so many Buick dealers that each is on the brink of financial ruin. The unfortunate consequence of this local monopoly power is that each distributor will maximize its profit by restricting output below the level that the manufacturer finds optimal.

Suppose that the manufacturer sells his output in a national market through a system of distributors to final consumers. Due to the nature of the distribution function, only one distributor will be established in each geographic submarket. The retail demand for the product in a given submarket is represented by D_R in Figure 3.2 along with the associated marginal revenue as MR_R. We have denoted the marginal cost of retailing as MC_R. If the retail distribution were competitively organized, the curve labelled $D_R - MC_R$ would be the derived demand. The curve marginal to $D_R - MC_R$ is labelled d, which equals $MR_R - MC_R$.

Since the retail distributor is a local monopolist, he or she will maximize profits by equating marginal revenue to marginal cost. The curve labelled $d = MR_R - MC_R$ is marginal revenue minus the marginal cost of retailing. For the distributor, marginal cost is the sum of the price charged by the manufacturer plus the marginal cost of retailing. Thus, the distributor will select the profit-maximizing output by equating the price he or she has to pay to the manufacturer with the net marginal revenue $MR_R - MC_R$. Consequently, when the distributor has a local monopoly, the effective derived demand for the submarket is d.

Manufacturers exploit their monopoly power by the selection of price and output at the point at which their marginal cost and marginal revenue are equal. In Figure 3.2, the marginal revenue for d is labelled mr. In

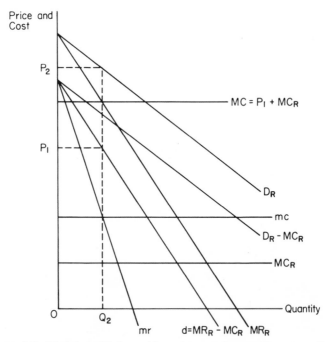

Figure 3.2 Market equilibrium with nonintegrated successive monopolies.

addition, the manufacturer's marginal production costs are mc. The manufacturer will produce Q_2 units of output for this market and will charge P_1 per unit. This price and output generates profit for the manufacturer of $(P_1 - mc)Q_2$.

The distributor will have a marginal cost MC equal to the price paid to the manufacturer P_1 plus the marginal cost of retailing MC_R. The distributor with market power maximizes its profit by equating this marginal cost $(P_1 + MC_R)$ to its marginal revenue (MR_R). Consequently, the distributor will sell Q_2 units of output to retail customers at a retail price of P_2. The distributor earns excess profits of $(P_2 - MC)Q_2$. Thus, the distributor benefits from his or her status as a local monopolist.

In order to see the adverse effects of successive monopoly on the manufacturer's profits, we have to compare this case with the situation in which distribution is competitive. In Figure 3.3, we have combined the results of Figures 3.1 and 3.2. We can see that the result of successive monopoly at the production and distribution stages is to reduce output from Q_1 to Q_2 and to raise the price to the consumer from P_R to P_2. With constant costs and linear demand, the intermediate product price remains at P_1 regardless of the market structure at the downstream stage. (See

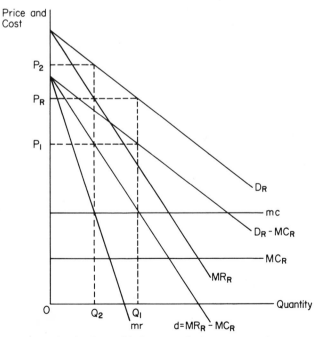

Figure 3.3 Reduction in total industry profit due to successive monopoly.

Greenhut and Ohta, 1976. Also, see the comments by Haring and Kaserman, 1978, and Perry, 1978, and the reply by Greenhut and Ohta, 1978.) The manufacturer's excess profits are reduced from $(P_1 - mc)Q_1$ to $(P_1 - mc)Q_2$. At the same time, the distributors' excess profits go from 0 to $(P_2 - MC)Q_2$. Thus, we have the following result (which is the result of Spengler, 1950, and Machlup and Taber, 1960):

> *Proposition 3.1*   The total industry profits are lower with successive monopoly than with a monopoly of manufacturing and competitive distribution.

Proof   With linear final good demand and constant costs at both stages of production, it is easily shown that $Q_1 = 2Q_2$, as we have drawn in Figure 3.3.[1] Given this result, it is then easy to demonstrate that industry profits are lower with monopoly at both stages. To do this requires showing that

$$(P_1 - mc)Q_1 > (P_1 - mc)Q_2 + (P_2 - MC)Q_2 \tag{3.1}$$

that is, upstream monopoly profits with competition at the downstream stage exceed the sum of monopoly profits at both stages with successive monopoly. Rearranging (3.1) and noting that $Q_1 = 2Q_2$, we have

$$(P_1 - mc)Q_2 > (P_2 - MC)Q_2$$

or

$$P_1 - mc > P_2 - MC \tag{3.2}$$

Now, profit maximization requires that

$$P_1 - mc = -Q_1 \frac{dP}{dQ} |_{D_R - MC_R}$$

where $dP/dQ|_{D_R - MC_R}$ is the slope of the inverse demand curve $D_R - MC_R$. Also, profit maximization requires that

$$P_2 - MC = -Q_2 \, dP/dQ|_{D_R},$$

where $dP/dQ|_{D_R}$ is the slope of the inverse demand curve D_R. Substituting these into (3.2) and noting that

$$\frac{dP}{dQ}|_{D_R - MC_R} = \frac{dP}{dQ}|_{D_R},$$

[1] If $P_R = a - bQ$, then $Q_1 = (a - mc - MC_R)/2b$ and $Q_2 = (a - mc - MC_R)/4b$.

we have

$$Q_1 > Q_2 \qquad\qquad (3.3)$$

and the proof is complete.

Vertical Integration

It should be fairly obvious that the manufacturer will resent having his or her profit reduced below the maximum potential profit. One way around this is to vertically integrate forward. Since total industry profits will be increased by such integration, both the upstream and the downstream firms can be ameliorated.[2] All that is required is that, following integration, one of the two divisions behave as if it were in a competitive industry. Which division adopts a competitive posture does not matter; as long as one division sets price equal to its marginal cost and the other division equates marginal revenue to marginal cost, total industry profits will be maximized.

Contractual Alternatives

There is an obvious contractual alternative to ownership integration, namely, setting *maximum* resale prices. If the manufacturer establishes a maximum resale price of P_R, the distributor's marginal revenue curve becomes equal to P_R for all outputs between 0 and Q_1. This will prevent the distributor from restricting output below Q_1 because the distributor's marginal cost will now equal marginal revenue at Q_1 units of output. Fixing maximum resale prices restores the price and the quantity that would result from competition at the distribution stage. For the retail customer, price is lower and a larger quantity is consumed when the manufacturer sets maximum resale prices. Thus, the manufacturer's interests are consistent with those of the consumers. In this case, the manufacturer's pursuit of profit benefits the consumer.

We should recognize that these maximum resale prices usually come as no surprise to the distributor. In the newspaper case, for example, one of the terms of the usual franchise agreement is that the distributor is expected to charge no more than the price that the publisher advertises for home delivery. Consequently, most distributors cannot legitimately claim to be surprised that pricing discretion is curtailed. In other words, the rules of the game are not changed after a large financial commitment has

---

[2] With linear demand and constant costs, the right-hand side of equation (3.1) will be exactly three-fourths of the left-hand side. Thus, total industry profits will go up one-third with vertical integration.

been made. Thus, there is no serious equity question that requires resolution.

Fixing maximum resale prices is also an effective way of dealing with a price-fixing conspiracy at the retail level. If the ostensible competitors at the retail distribution stage conspire to raise prices, then the effect is analogous to the successive monopoly case. One way of thwarting such efforts is to impose maximum resale prices. For a case in which this was the operative motivation, see *Kiefer-Stewart Co.* v. *Joseph E. Seagram & Sons* 340 U.S. 211 (1951).

In principle, the simple model developed above may also be applied to the more complex case of multiple products. Many retail food franchises, for example, McDonald's, Burger King, and Kentucky Fried Chicken, involve multiple products. The success of these operations depends inter alia upon maintaining fairly uniform quality standards across franchisees. In addition, relative prices must be maintained vis-a-vis competitive products. Neither the franchisor nor the franchisees benefit from any weakening in consumer confidence that may result when buyers are surprised at prices that are out of line with their past experience. Since such pricing can occur when independent franchisees that obtained prime locations attempt to extract higher prices, the franchisor has an interest in the prices charged by each franchisee. This interest springs from two sources: first, the successive monopoly problem discussed above and, second, the spillover effect that one franchisee's pricing behavior can have on the sales of other franchisees.

An additional contractual alternative to ownership integration in the successive monopoly situation is the stipulation of performance standards. If the upstream monopolist establishes a minimum quantity of sales equal to Q_1, then the distributors will be unable to restrict their output. Thus, appropriately selected performance standards may be used to achieve results that are equivalent to both ownership integration and maximum resale prices.

Important Product-Specific Services

The following section relies on the perceptive analyses provided by Bowman (1955) and Telser (1960). Whenever a product is sold, the customer jointly buys a package of services. These services include the amenities of the sales outlet, the seller's location, the provision of credit, the number of sales people per customer, the hours of operation, and so on. All of these services pertain to the way that a seller does business generally. We should expect these services to vary across sellers, thereby permitting a seller to cater to the tastes of a particular group of customers. In other words, there will be many different retailers offering different combinations of services in conjunction with any particular commodity. As a consequence, the cost of providing the commodity plus the services will vary across sellers. Thus, we should expect competition to result in an array of prices rather than a single price.

Ordinarily, the manufacturer of a commodity is largely unconcerned with the business practices followed by the distributors of his or her output. Competition will eliminate the inefficient and the unscrupulous. But this lack of concern extends only to the distributor's general business practices. It does not extend to services that are product-specific. There are numerous examples of product-specific services: the seller of complicated consumer goods like cameras or stereo equipment must explain the technical features to prospective customers; the seller of automobiles provides test drives for potential buyers. These kinds of services are important to the manufacturers because they affect the demand for his product. It is not hard to see that if no seller would provide test drives, then the demand for automobiles would be lower than if test drives were provided. This follows from the fact that the information contained in a test drive is valuable to the customer. The nature of this problem can be seen in Figure 3.4.

The retail demand for the commodity in question is denoted by D if no product-specific services are provided. The marginal (and average) costs of retailing are given by MC_D. Consequently, the manufacturer's derived demand is $d = D - MC_D$, and the associated marginal revenue is mr. Given the manufacturer's marginal cost of production, MC_P, the optimal price and quantity for the manufacturer are p and q, respectively. The retail price will be P, which is equal to p plus MC_D. Consequently, the competitive retail distributors will earn 0 excess profits. The manufacturer's profits will be $(p - MC_P)q$.

This outcome can be compared to the results when the product-specific services are provided. The provision of the services shifts the retail demand curve to D^*. Since the additional services are costly, the marginal (and average) distribution costs rise to MC_D^*. The vertical shift in final product demand must exceed the increase in the marginal distribution costs. Otherwise, there would be no profit incentive to provide these services at all. Now, the manufacturer's derived demand curve becomes $d^* = D^* - MC_D^*$ and the associated marginal revenue curve is mr*. The manufacturer obviously benefits from this shift in demand. His or her new optimal price and quantity are higher: p^* and q^*. The retailers will then charge a higher price equal to p^* plus MC_D^*. This higher price does not

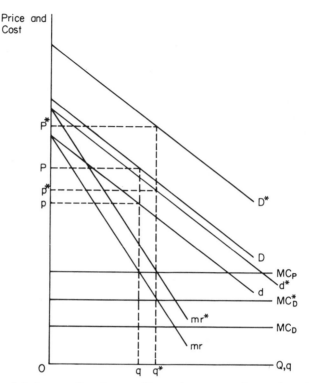

Figure 3.4   Impact of product-specific services on manufacturer's profit.

yield any excess profits for the distributors. In contrast, the manufacturer's profit expands to $(p^* - MC_p)q^*$. Consequently, the manufacturer wants these services provided to the retail customers.

Vertical Integration

One way of being sure that the product-specific services are provided is for the manufacturer to vertically integrate forward into distribution. In some cases, this is a viable option. For example, when other avenues to vertical control were foreclosed, Schwinn began to distribute its bicycles through its own retail outlets rather than through independent franchisees. In *United States* v. *Arnold, Schwinn & Co.*, 388 U.S. 365 (1967), Schwinn's restrictions on its franchisees were held to violate the antitrust laws per se. In principle, automobile manufacturers could also own all of their dealerships. But there are other products that can benefit from product-specific services where vertical integration is not a sensible option.

For example, the manufacturer of a particularly sophisticated piece of stereo equipment *could* vertically integrate forward. But he or she would have to carry a full complement of products or there would be little customer traffic. Thus, the manufacturer may be able to specialize in production, but be unable to do so at the distribution stage. This may make vertical integration an unrealistic practical option although it remains a theoretical alternative. There are, however, practical alternatives to ownership integration.

Contractual Alternatives

Rather than vertically integrating forward, the manufacturer may ask each of its distributors to provide the desirable product-specific services. If each of them honors this request, the demand shifts to D^* and the results described above come to pass. The problem with product-specific services is that one can consume them separately from the product itself. Suppose that all distributors are offering the requested services. Each is incurring costs of MC_D^* per unit and selling the product for P^*. None of the distributors is earning any excess profit because the difference between P^* and MC_D^* equals the price charged by the manufacturer. Sooner or later, it will occur to one of the distributors that profits can be improved by not offering some or any of the product-specific services. As long as the rest of the distributors continue providing those services, the final-good demand will remain at D^*. By cutting out some services, the clever distributor can cut price just below P^*, thereby increasing the volume of business. Thus, the clever distributor becomes a "free rider" who benefits from the expenditures of other distributors without having to make comparable expenditures. In addition, profits will soar as costs are considerably reduced. For example, suppose a discount automobile dealer offered no services. Its costs would fall to MC_D from MC_D^* while it need only offer a price marginally lower than P^* to attract many customers. Thus, one dealer can convince the customer to buy while another dealer makes the sale by offering a lower price.

This becomes a real problem when a lot of distributors follow suit because the demand curve will begin to drift downward from D^* toward D. The excess profits of the early defectors will prove to be transitory as competition leads to lower and lower prices. The final equilibrium at which none of the product-specific services are offered has been described above. The social welfare effects of free riders forcing the industry to this equilibrium are theoretically indeterminate. Since the provision of product-specific services involves the expenditure of resources on an activity that shifts demand, the welfare analysis of advertising introduced by Dixit and Norman (1978) applies.

This analysis requires adopting either the preadvertising or the postadvertising demand (here, pre- or post-product-specific services demand) as the standard for evaluating the change in output associated with a movement from the former equilibrium to the latter. Since output is increased by the provision of product-specific services, this valuation will be positive regardless of which standard is adopted. Which demand curve one selects as the standard depends upon whether the product-specific services serve an informative or a deceptive role. If these services are informative, then the postservice demand will represent the true preferences of the consumer. If they are deceptive, however, the preservice demand represents the true preferences. (See Dixit and Norman, 1978, pp. 1–2.)

To determine the net effect of providing these services on overall social welfare, however, we must subtract the incremental costs of producing the additional output plus the resource costs of providing the product-specific services from this valuation of the output change. In Figure 3.4, these two categories of costs correspond to the area under MC_P from q to q^* and the area between MC_D and MC_D^* from 0 to q^*, respectively. As shown by Dixit and Norman (1978), the net effect may be positive or negative regardless of which demand curve is selected as the standard. Thus, we are unable to determine whether society as a whole is better off with or without the product-specific services. Dixit and Norman (1978, p. 5) demonstrate that the equilibrium quantity of advertising selected by a monopolist will be excessive from a social welfare point of view. This result would appear to carry over to the provision of product-specific services. Despite this excess, however, we cannot conclude that a movement to 0 product-specific services would result in an improvement in welfare. It is clear, however, that upstream monopolists are better off if they can encourage their distributors to provide these services.

Resale price maintenance can be used to prevent the free rider problem. If the manufacturer specifies that all units of the product must be sold at a price of P^* or more, there will be little incentive to stop providing the services. It is true that a single dealer can reduce costs by reducing or eliminating the services. But, with price fixed at P^*, there is no way to attract the customer. Thus, a reduction in services will result in lost sales and probably in negative profits. In the absence of price competition, the scramble for retail sales will lead to nonprice competition, which is not susceptible to free-rider problems. The retailers will be forced to provide the level of services that leads to maximum profits for the manufacturer.

If a minimum resale price is specified by the manufacturer, a recalcitrant distributor can still *attempt* to be a free rider. In other words, the resale price-maintenance agreement will have to be policed. This, of course, is true of many contracts. As long as the costs of policing the agreement are smaller than the difference in the profit levels,

$$\Delta\pi = (p^* - MC_p)q^* - (p - MC_p)q$$

it will be sensible to impose a resale price-maintenance agreement. There exist, however, other contractual alternatives to resale price maintenance.

There is a host of nonprice vertical restraints that are usually referred to as territorial and customer restrictions. For example, there are exclusive territories, areas of primary responsibility, profit passovers, location clauses, and so on. It can be shown quite easily that these restrictions are alternatives (albeit imperfect ones) to resale price maintenance. Bork (1966a) provided a thorough development of market division as an alternative to vertical price fixing or resale price maintenance. In addition, he explains why some variants, for example, profit passovers and areas of primary responsibility, are rather imperfect substitutes. It is evident that market division and customer allocation schemes tend to accomplish the same objective. By assigning each geographic market to a specific distributor, a manufacturer can remove the incentives for price competition among his distributors. Similarly, by assigning each customer or class of customers to a specific distributor, there is also no incentive for price competition. This does not preclude the manufacturer from performing part of the distribution function. In fact, such was the case in *White Motor* where White reserved certain customers for direct sales. In effect, the manufacturer creates a successive monopoly situation. By itself, this development would not be in the manufacturer's interest. Each distributor will have a local monopoly at the retail level. Consequently, each distributor will provide the quantity of commodity-specific services that maximize profits. But the manufacturer is left with the problem of pushing the distributors to provide the optimal level of services from the manufacturer's point of view.

One way of forcing the distributor to provide more extensive services is through the imposition of quotas or other performance standards. Given that the economic value of the distributorship remains positive, the threat of termination should be sufficient to induce the distributors to expand sales beyond the distributor's profit maximizing level. In other words, the manufacturer can force an expenditure of resources on sales-promoting services that will eliminate monopoly profits at the retail stage. This will approximate the resale price maintenance results. When the physical commodity involves trade-ins, market division may be superior to resale price maintenance. In fact, resale price maintenance will be ineffective in the presence of trade-ins, but market division eliminates retail competition more completely and will be more effective.

Entry Barriers

Later, in Chapter 6, we shall show that vertical integration may, at times, be employed as a method to overcome existing barriers to entry. In this section, however, we examine the more traditional view of the relationship between vertical integration and entry barriers, namely, the possibility that vertical integration may be used as an instrument for increasing the impediments to entry in a given industry.

Debate concerning the potential impact of vertical integration on barriers to entry has continued for a long time. One side of this debate (e.g., Bork, 1969 and 1978) argues that the presence or absence of vertical integration cannot affect the difficulty of entering an industry. The other side (e.g., Comanor, 1967 and Williamson, 1979) argues that vertical integration by existing firms may increase the entry barriers faced by potential new firms. With no expectation of completely resolving this debate, we shall attempt here to outline the major issues involved. Our discussion is drawn from Kaserman (1978).

Potential Impact

To begin, if the potential relationship between vertical integration and barriers to entry is to have any policy interest, it is clear that some market power must be present at one or more of the vertically related stages of production. Potential competition from new entrants assumes importance only where actual competition among the existing members of the industry is absent or, at least, imperfect. Thus, as Williamson (1979, p. 962) pointed out, this topic is of interest only in dominant firm industries or collusive oligopolies.

Next, if vertical integration is to exert an influence on entry barriers, new firms must feel compelled to enter more than one stage simultaneously or, at least, suffer some cost disadvantage from not doing so. That much is tautological; for, if potential entrants could assume production status at one stage only without penalty, then the vertical structure of the existing firms would not affect the conditions of entry. Multistage entry may be deemed necessary for two potential reasons. First, if the industry is completely vertically integrated in the sense that all existing firms manufacture their entire requirements of the intermediate product and transfer their entire output of this product internally, then the input market will not exist, and single-stage entry will be completely forestalled unless the potential entrant can expect simultaneous independent entry at the related stage or stages. The degree of coordination between the potential entrants that would be required to achieve such simultaneous entry makes this a highly unlikely event. Similarly if vertical integration is incomplete but pervasive, then the size of the intermediate product market may be severely constrained (Edwards, 1953, pp. 407–408). If so, economies of scale or indivisibilities in the production of the intermediate good may inhibit independent single-stage entry; nonintegrated firms may suffer bargaining difficulties, particularly if they are forced to transact with integrated firms; and the threat of being subjected to rationing will be increased if the intermediate product market should fail to clear. Many of these potential difficulties of single-stage entry in the presence of pervasive vertical integration revolve around what Williamson (1974) refers to as small-numbers bargaining problems. As described in Chapter 2, transaction costs are apt to be high in cases in which the number of potential trading partners is limited. Here, however, it is the presence of pervasive vertical integration that gives rise to significant transaction costs. In other words, the direction of causation may be the reverse of that described earlier.

Second, if production costs are lower for vertically integrated firms, either because of transaction cost savings or the ability to circumvent noncompetitive pricing in the input market, then potential entrants may feel obliged to undertake entry at multiple stages simultaneously. Otherwise, the extra costs associated with single-stage entry may prohibit the achievement of a competitive return.

If, for any of these reasons, new firms are forced to enter at more than one stage of production, then vertical integration can conceivably exert an influence on entry barriers. Such an influence may emanate from two basic sources. First, if some barrier to entry exists at one stage but is absent at another, then the necessity of multi-stage entry will transmit this barrier from the former to the latter.[3] Thus, integrated entry can be expected to be at least as difficult as single-stage entry at the stage with the highest barriers.

Second, a requirement that entry occur simultaneously at more than one stage necessarily increases the capital needs of potential entrants. Such an increase may or may not impede entry, depending on the existence of an imperfection in the capital market (defined as a situation in which the terms of finance deteriorate with an increase in the quantity of funds raised). Smith (1971) has shown that lender risk-aversion and a

---

[3] Difficulties in developing technological expertise at one or more of the production stages fall into this category. See Stigler (1968, p. 191), and Williamson (1974).

positive probability of borrower default will yield such a result if funds are raised by debt flotation. Also, Williamson (1971) argued that difficulties of monitoring the performance of large integrated firms may result in higher returns being required by investors if funds are raised in the equity market. Therefore, it is entirely plausible that vertical integration impedes entry by increasing the capital requirements of potential entrants. Whether or not the degree of vertical integration exhibited by the existing members of an industry actually does exert a positive effect on the per-unit capital costs of new entrants would appear to be an empirical question that is amenable to investigation, especially since the appearance of Maddigan's (1981) operational measure of the degree of vertical integration.

Optimal Integration

On the basis of the preceding arguments, it does not appear unreasonable to expect vertical integration by existing firms to lead to an increase in the barriers to entry faced by potential new firms. If such an effect does exist, then we should expect some firms to engage in vertical integration in situations in which current profits are reduced by such action, that is, where, due to managerial diseconomies or other reasons, costs are increased by internalization of the additional stage of production. Comanor (1967) has characterized such behavior as an "investment in entry barriers." (Also, see Williamson, 1979, ft. 45, p. 962.) A simple model in the spirit of that presented by Spence (1977, pp. 542–543) may be used to illustrate the incentives behind this strategy.

Assume that we have a nonintegrated intermediate-product monopolist that may decide to increase the barriers to entry into its industry by engaging in forward integration. We assume that this firm operates in two periods, which we label the current period and the future period. Entry can occur in the future period only. Then, to examine the decision problem faced by this firm in selecting the optimal level of vertical integration, we adopt the following notation and assumptions:

z = the fraction of the productive capacity at the downstream stage that is integrated in the current period;

$p(z)$ = the probability that entry will be forestalled in the future period as a function of the degree of vertical integration adopted in the current period, where we assume $dp/dz > 0$;

$\pi_1(z)$ = current period profits, which we assume may be either increased or decreased by vertical integration, that is, $d\pi_1/dz \lessgtr 0$

π_2^{NE} = future period profits if entry does not occur; and

π_2^{E} = future period profits if entry does occur, where $\pi_2^{E} < \pi_2^{NE}$.

Given this notation, the expectation of the present value of the firm's profits is given by

$$\bar{\pi} = \pi_1(z) + \frac{1}{1 + r} \{p(z)\pi_2^{NE} + [1 - p(z)]\pi_2^{E}\} \tag{3.4}$$

where r is the firm's discount rate. The intermediate product monopolist attempts to maximize $\bar{\pi}$ subject to the constraint that $0 \leq z \leq 1$.

The Lagrangian for this problem is

$$L = \pi_1(z) + \frac{1}{1 + r} \{p(z)\pi_2^{NE} + [1 - p(z)]\pi_2^{E}\} + \lambda(1 - z)$$

with Kuhn-Tucker conditions:

$$\frac{d\pi_1}{dz} + \frac{1}{1 + r} \frac{dp}{dz} (\pi_2^{NE} - \pi_2^{E}) - \lambda \leq 0 \tag{3.5}$$

$$z \geq 0 \tag{3.6}$$

$$\left[\frac{d\pi_1}{dz} + \frac{1}{1 + r} \frac{dp}{dz} (\pi_2^{NE} - \pi_2^{E}) - \lambda\right]z = 0 \tag{3.7}$$

$$1 - z \geq 0 \tag{3.8}$$

$$\lambda \geq 0 \tag{3.9}$$

$$\lambda(1 - z) = 0 \tag{3.10}$$

From these conditions, we may identify three cases for discussion.

Case 1: $\lambda = 0, z = 0$ If $z = 0$, that is, if it is optimal for the firm to remain completely nonintegrated, then condition (3.10) implies $\lambda = 0$ as well. Then, condition (3.5) implies that, in this case,

$$-\frac{d\pi_1}{dz} \geq \frac{1}{1 + r} \frac{dp}{dz} (\pi_2^{NE} - \pi_2^{E}) \tag{3.11}$$

Since the right-hand side of (3.11) is positive, it is necessary that $d\pi_1/dz < 0$ hold for all values of z. Intuitively, the loss in current period profits that results from the first unit must be large enough to equal or exceed the gain in expected discounted future period profits for the firm to select this option. In other words, $d\pi_1/dz$ must be both negative and large even for low values of z.

Case 2: $\lambda = 0, 0 < z < 1$ Here, the firm finds it optimal to engage in partial vertical integration. With $z < 1$, condition (3.10) implies $\lambda = 0$. For this case to result, conditions (3.5) and (3.7) imply that, at the optimal degree of vertical integration,

$$-\frac{d\pi_1}{dz} = \frac{1}{1+r}\frac{dp}{dz}(\pi_2^{NE} - \pi_2^{E}) \tag{3.12}$$

must hold. Since the right-hand side of (3.12) is positive, it is again necessary that $d\pi_1/dz < 0$ hold at the solution value. Here, however, the firm chooses to engage in some forward integration in the current period. Such integration is pursued up to the point at which the marginal loss in current period profits is just equal to the marginal gain in expected discounted future period profits resulting from the increased probability that entry will be forestalled. This might be the most likely case in the event that $d\pi_1/dz > 0$ for low values of z but $d\pi_1/dz < 0$ for increasingly larger values of z.

   Case 3: $\lambda > 0$, $z = 1$ This is the case of complete vertical integration. In this case, conditions (3.5) and (3.7) imply that

$$\frac{d\pi_1}{dz} + \frac{1}{1+r}\frac{dp}{dz}(\pi_2^{NE} - \pi_2^{E}) = \lambda > 0 \tag{3.13}$$

This case will necessarily result if $d\pi_1/dz > 0$ holds for all values of z. Here, vertical integration increases both current period profits and expected discounted future period profits. Consequently, there is no reason for the firm to stop short of complete integration. This same case could, of course, result if $d\pi_1/dz < 0$. Here, however, the second term on the left-hand side of (3.13) would have to exceed $d\pi_1/dz$ in absolute value. In other words, the expected gain from reducing the probability of entry would have to exceed the loss in current period profits for all values of z.

   Finally, in the event that vertical integration has no effect on the probability of entry being forestalled in the future period, we set $dp/dz = 0$ in the above conditions. Then, it is easily shown that the three cases described above of zero, partial, or complete vertical integration will emerge depending upon the values of $d\pi_1/dz$ over the different values of z. That is, the decision to integrate during the current period will hinge, as one might suspect, entirely upon the effect of such integration on current period profits.

A Contractual Alternative

   In cases 1 and 2 above, the firm found it optimal to remain less than completely vertically integrated because of the negative effect of integration on current period profits. Assuming that further integration would lead to an additional increase in the probability that entry will be forestalled (i.e., that $dp/dz > 0$), the firm may seek other avenues to retard entry. Williamson (1979, p. 965) argued that exclusive dealing contracts may be employed to achieve this end. Spence (1977, p. 544) also made a brief mention of these contractual arrangements as a method for impeding entry. By tying a given distributor to a given manufacturer's product, such contracts may force new firms to enter both stages simultaneously, thereby raising the capital requirements for successful entry.[4] As Goldberg (1979, ft. 96, p. 114) pointed out, however, exclusive dealing arrangements are not likely to provide results that are completely equivalent to ownership integration because the exit barriers faced by distributors are likely to be less severe.[5]

[4] Note, that, in case 2, the firm employs partial vertical integration and, additionally, a system of exclusive dealerships. This case, then, corresponds to the observed practice of dual distributorships.

[5] An excellent discussion of exit barriers and their role in structuring what Goldberg calls "relational contracts" may be found in Goldberg (1979).

4

Variable Proportions and Contractual Alternatives

In the preceding chapter, we were concerned with the incentives for vertical control by an upstream monopolist whose output was employed in fixed proportions by the downstream industry. Now, we shall turn our attention to the case of variable proportions. We shall show that when the downstream industry is able to employ the monopolized input in variable proportions, the intermediate-product monopolist has an incentive to vertically integrate forward. First, we shall establish this result for ownership integration. Subsequently, we shall examine four contractual alternatives to vertical ownership integration. Specifically, we shall discuss tying arrangements, output royalties, sales-revenue royalties, and entry fees as contractual alternatives in response to variable proportions production.

Ownership Integration

In an important, albeit somewhat neglected, article, Meyer Burstein (1960) pointed out that an input monopolist has an incentive for forward integration when the downstream industry is able to employ the monopolized input in variable proportions. Burstein's contribution has been rediscovered, formalized, and extended in a series of articles dealing with what has come to be known as the variable proportions incentive for vertical integration. A heuristic demonstration of this incentive was provided by Vernon and Graham (1971).

Consider the case in which the final good is produced according to a production function that combines inputs in variable proportions. Suppose that input x_1 is produced by a monopolist while input x_2 is supplied by a competitive industry. When the x_1 monopolist attempts to maximize

48

profit, the resulting price of x_1 leads the final-good producers to substitute away from x_1 in favor of x_2. This behavior creates an incentive for the x_1 monopolist to vertically integrate forward.

In Figure 4.1, we have drawn the industry isoquant for the total output produced by the competitive final-good industry.[1] Thus, the isoquant labeled Q_O shows all of the input combinations that could be used to produce the industry output. Given the competitive price of input x_2 and the monopolistic price of x_1, the final-good industry selects the input combination at point Z where isocost AB is tangent to the Q_O isoquant. The slope of AB reflects the ratio of the input price p_1/MC_2, where MC_2 is the marginal cost. Since x_2 is produced competitively, its price will be equal to marginal cost. As a result, the equation of the isocost is

$$x_2 = \frac{M}{MC_2} - \frac{p_1}{MC_2}x_1$$

where M is the dollars spent, MC_2 the marginal cost of producing input x_2, and p_1 the profit maximizing price of input x_1. The intercept at A tells us how many units of x_2 one can buy with M dollars if no x_1 is purchased.

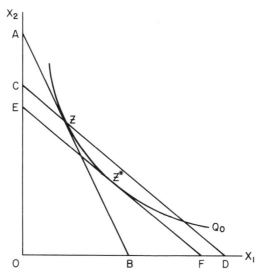

Figure 4.1 Incentive to integrate due to variable proportions at the downstream stage. From Vernon, John, and Graham, Daniel. "Profitability of Monopolization by Vertical Integration." *Journal of Political Economy*, Vol. 79 (July/August 1971), pp. 924–925. © by The University of Chicago. All rights reserved.

[1] Waterson (1982) explores a variable proportions model in which the downstream industry is oligopolistic.

Consequently, we can measure M in terms of the units of x_2. In this case, the combination at Z costs A units of x_2 to produce.

The isocost CD goes through Z, but has a slope reflecting the ratio of competitive prices for x_1 and x_2: MC_1/MC_2. Now, at competitive prices for both inputs, we can see that an expenditure equal to C units of x_2 must be made to produce Q_0 units of the final good with the input combination at Z. Consequently, the x_1 monopolist enjoys a profit equal to the difference between what input combination Z costs when x_1 is priced monopolistically and what it costs when x_1 is priced competitively. In the graph, this is equal to $A - C$ units of x_2.

Suppose that the x_1 monopolist vertically integrated forward by buying all of the final-good producers. The input monopolist could increase profits simply by using a more efficient input combination and would not produce Q_0 by using the inefficient combination at Z. This combination resulted from the competitive producers substituting away from the relatively higher priced input x_1. In order to produce Q_0 efficiently, the vertically integrated firm would have to minimize the costs of producing Q_0. This requires employing the input combination at Z^* where the EF isocost is tangent to the Q_0 isoquant. In terms of units of x_2, the monopolist's profit has increased from $A - C$ to $A - E$. This increase in profit provides the incentive for vertical ownership integration when the final good is produced according to variable proportions. This analysis assumes that the output remains at Q_0 after the vertical integration. This, of course, may not be the case. If optimal output changes, the new equilibrium must involve an even larger increase in profit because the firm always has the option of staying at Q_0.

A More Formal Proof We can establish this incentive for forward integration in a more formal fashion. If the input monopolist is not vertically integrated, there will be an upstream profit function of

$$\pi^U = p_1(x_1)x_1 - c_1x_1 \tag{4.1}$$

where p_1, x_1, and c_1 are the price, quantity, and average (and marginal) production cost, respectively, of the monopolized input. The profit function of the downstream final-good producers can be written as

$$\pi^D = PQ - p_1x_1 - c_2x_2 \tag{4.2}$$

where P and Q are the price and output of the final good while c_2 and x_2 are the competitive price and quantity of the second input.

The first-order conditions for a maximum of (4.2) are

$$\frac{\partial \pi^D}{\partial x_1} = P\frac{\partial Q}{\partial x_1} - p_1 = 0 \tag{4.3}$$

and

$$\frac{\partial \pi^D}{\partial x_2} = P \frac{\partial Q}{\partial x_2} - c_2 = 0 \tag{4.4}$$

Solving (4.4) for P and substituting into (4.3) yields

$$p_1 = c_2 \frac{\partial Q/\partial x_1}{\partial Q/\partial x_2} \tag{4.5}$$

Substituting (4.5) into (4.1) provides an expression for the upstream monopolist's profit when the optimizing behavior of the downstream firms is taken into account:

$$\pi^U = \left(\frac{c_2}{\partial Q/\partial x_2} \frac{\partial Q}{\partial x_1}\right) x_1 - c_1 x_1 \tag{4.6}$$

If the upstream monopolist vertically integrates forward, his profit function would be

$$\pi_I = PQ - c_1 x_1 - c_2 x_2 \tag{4.7}$$

Now, we can compare π^U and π_I. Dropping $c_1 x_1$ from (4.6) and (4.7), we want to determine if

$$c_2 \frac{\partial Q/\partial x_1}{\partial Q/\partial x_2} x_1 < PQ - c_2 x_2 \tag{4.8}$$

Adding $c_2 x_2$ to both sides of (4.8), we have

$$c_2 \left(\frac{\partial Q/\partial x_1}{\partial Q/\partial x_2} x_1 + x_2\right) < PQ$$

or

$$c_2 \left[\frac{(\partial Q/\partial x_1)x_1 + (\partial Q/\partial x_2)x_2}{\partial Q/\partial x_2}\right] < PQ \tag{4.9}$$

By Euler's theorem, the numerator of the term in brackets is Q. Thus, we have

$$\frac{c_2}{\partial Q/\partial x_2} Q < PQ$$

or

$$\frac{c_2}{\partial Q/\partial x_2} < P \tag{4.10}$$

Since the left-hand side is marginal cost, we have established the inequality because the profit-maximizing firm will equate marginal cost and marginal revenue, which is necessarily less than price.

Following the Vernon and Graham proof that variable proportions provide an incentive for forward integration, Schmalensee (1973) has shown that the incentive to integrate downstream persists until the intermediate-good monopolist has successfully monopolized the final-good industry. Thus, a potential incentive for extension of the monopoly to later stages of production (possibly facilitated by a price-cost squeeze of nonintegrated downstream producers) has been established. Also, Hay (1973), Warren-Boulton (1974), Mallela and Nahata (1980), and Westfield (1981) have demonstrated that, given the existence of monopoly power at the upstream stage, the social welfare effects of a successful monopolization of the downstream industry are a priori indeterminate. The direction of the welfare effect of an extension of monopoly power under these conditions has been shown to depend on the specific production and demand conditions present. We shall return to this issue when we discuss the public policy implications of our analysis.

Tying Arrangements

The original Burstein (1960) paper argued that, as an alternative to vertical integration, the intermediate-good monopolist could obtain an equivalent result by tying the purchase of nonmonopolized substitutable inputs to the purchase of the monopolized input. Thus, Burstein argued that the upstream firm could, through tying, exercise the requisite vertical control necessary to achieve the same result as that available through the successful monopolization of the downstream industry by ownership integration. Burstein, however, did not *prove* that tie-in sales would lead to equivalent results. Nonetheless, his assertion can be proved. (This proof was offered first by Blair and Kaserman, 1978a).

Let x_1 and x_2 be inputs in the production of final output Q. Let c_1 and c_2 be the constant marginal costs of inputs x_1 and x_2, respectively. Finally, let $Q(x_1, x_2)$ represent the final-good industry production function, which we assume to be linearly homogeneous in x_1 and x_2.[2] In addition, suppose

---

[2] In essence, we are assuming that the long-run supply curve is horizontal when each competitive firm produces in a single plant. Alternatively, if all plants were under the control of one management, then the long-run marginal cost of a multiplant operation would be constant. This, of course, requires the absence of interplant economies or diseconomies. Returns to scale for each plant will be locally constant along this long-run marginal cost curve. See Patinkin (1947).

that the production of x_1 is monopolized while the markets for x_2 and Q are competitive. If the input monopolist had integrated forward, its profit function would be given by

$$\pi_I = P[Q(x_1, x_2)]Q(x_1, x_2) - c_1 x_1 - c_2 x_2 \qquad (4.11)$$

where $P[Q(x_1, x_2)]$ is the inverse demand function for final output. The vertically integrated monopolist's behavior is then characterized by maximization of π_I over x_1 and x_2 with c_1, c_2, $P(Q)$, and $Q(x_1, x_2)$ taken as given. Under this strategy, the intermediate-product market is replaced by internal transfers and the downstream industry would be monopolized.

In contrast, suppose the input monopolist employs a tying arrangement. The upstream monopolist can purchase x_2 at the competitive price c_2 and tie the subsequent purchase of x_2 by competitive downstream producers to the purchase of the monopolized input x_1. (Of course, the monopolist could produce x_2 itself. There is, however, nothing to be gained by doing so, because our assumption of a competitive market for x_2 implies that it can be purchased at constant marginal cost c_2.) Following a tying strategy then, the monopolist's profit function is given by

$$\pi_T^U = p_1(x_1, x_2)x_1 + p_2(x_1, x_2)x_2 - c_1 x_1 - c_2 x_2 \qquad (4.12)$$

where $p_i(x_1, x_2)$ is the inverse derived demand function for the ith input, $i = 1, 2$. The monopolist's behavior is then characterized by the maximization of π_T^U over x_1 and x_2 with c_1, c_2, $p_1(x_1, x_2)$, and $p_2(x_1, x_2)$ taken as given.

Under this strategy, an intermediate-product market for x_1 will continue to exist, and the downstream industry will face the profit function

$$\pi_T^D = PQ(x_1, x_2) - p_1 x_1 - p_2 x_2 \qquad (4.13)$$

Firms in this industry will accept P, p_1, p_2, and $Q(x_1, x_2)$ as given and maximize π_T^D over x_1 and x_2.

The Equivalence Between Vertical Integration and Tying

Under the assumptions of the variable proportions model, vertical integration and tying are economically equivalent. That is, these strategies yield identical results with regard to both efficiency of input utilization and profitability to the intermediate-good monopolist. This assertion can be shown by proving the following two theorems (also found in Blair and Kaserman, 1978a):

Theorem 4.1 Given the assumptions of the variable proportions model, a successful monopolization of the downstream industry through vertical integration *and* a tying arrangement whereby the purchase of x_2 by downstream firms is tied to the purchase of x_1 both result in inputs x_1 and x_2 being combined in efficient proportions.

Proof Efficient production requires that input proportions be adjusted such that

$$\frac{\partial Q / \partial x_1}{\partial Q / \partial x_2} = \frac{c_1}{c_2} \tag{4.14}$$

that is, that the firm producing Q be on its expansion path. The necessary conditions for a maximization of π_I by the integrated monopolist are

$$\left(P + Q \frac{\partial P}{\partial Q} \right) \frac{\partial Q}{\partial x_1} = c_1 \tag{4.15}$$

and

$$\left(P + Q \frac{\partial P}{\partial Q} \right) \frac{\partial Q}{\partial x_2} = c_2 \tag{4.16}$$

Dividing (4.15) by (4.16) yields expression (4.14), thereby establishing the efficiency of input utilization under the vertical integration strategy. To establish the equivalent result under the tying arrangement, we note that profit maximization on the part of the downstream firms [maximization of π_T^D from expression (4.13)] requires that the value of the marginal product of each input be equated to its price, that is, that

$$P \frac{\partial Q}{\partial x_1} = p_1 \tag{4.17}$$

and

$$P \frac{\partial Q}{\partial x_2} = p_2 \tag{4.18}$$

Substituting (4.17) and (4.18) into (4.12) and differentiating with respect to x_1 and x_2, we obtain the first-order conditions for the maximization of the upstream monopolist's profits, π_T^U, taking into account the optimizing behavior of firms in the downstream industry:[3]

$$\left(P + \frac{\partial P}{\partial Q} \frac{\partial Q}{\partial x_1} x_1 + \frac{\partial P}{\partial Q} \frac{\partial Q}{\partial x_2} x_2 \right) \frac{\partial Q}{\partial x_1} + P \left(x_1 \frac{\partial^2 Q}{\partial x_1^2} + x_2 \frac{\partial^2 Q}{\partial x_2 \partial x_1} \right) = c_1 \tag{4.19}$$

---

[3] In taking these derivatives, we must now view final-output price P as a function of total downstream industry output Q in order to reflect the upstream monopolist's recognition of the negatively-sloped demand for the final product properly.

and

$$\left(P + \frac{\partial P}{\partial Q}\frac{\partial Q}{\partial x_1}x_1 + \frac{\partial P}{\partial Q}\frac{\partial Q}{\partial x_2}x_2\right)\frac{\partial Q}{\partial x_2} + P\left(x_2\frac{\partial^2 Q}{\partial x_2^2} + x_1\frac{\partial^2 Q}{\partial x_1 \partial x_2}\right) = c_2 \quad (4.20)$$

Linear homogeneity of the production function[4] implies that

$$x_1\frac{\partial^2 Q}{\partial x_1^2} + x_2\frac{\partial^2 Q}{\partial x_2 \partial x_1} = x_2\frac{\partial^2 Q}{\partial x_2^2} + x_1\frac{\partial^2 Q}{\partial x_1 \partial x_2} = 0$$

Thus, (4.19) and (4.20) reduce to

$$\left(P + \frac{\partial P}{\partial Q}\frac{\partial Q}{\partial x_1}x_1 + \frac{\partial P}{\partial Q}\frac{\partial Q}{\partial x_2}x_2\right)\frac{\partial Q}{\partial x_1} = c_1 \quad (4.21)$$

and

$$\left(P + \frac{\partial P}{\partial Q}\frac{\partial Q}{\partial x_1}x_1 + \frac{\partial P}{\partial Q}\frac{\partial Q}{\partial x_2}x_2\right)\frac{\partial Q}{\partial x_2} = c_2 \quad (4.22)$$

Division of (4.21) by (4.22) again results in expression (4.14), thereby establishing the efficiency of input proportions under the tying alternative and completing the proof of Theorem 4.1.

Theorem 4.2 Given the assumptions of the variable proportions model, a successful monopolization of the downstream industry through vertical integration *and* a tying arrangement whereby the purchase of x_2 by downstream firms is tied to the purchase of x_1 yield identical profits to the input monopolist.

Proof We want to show that the profit functions π_I and π_T^U, from expressions (4.11) and (4.12) respectively, are identical. Since the input costs appear identically through the cost equations, we may focus upon the revenue functions. Identical profits then require that

$$P[Q(x_1, x_2)]Q(x_1, x_2) = p_1(x_1, x_2)x_1 + p_2(x_1, x_2)x_2 \quad (4.23)$$

Substituting (4.17) and (4.18) on the right-hand side of (4.23) and factoring P, we obtain

$$P[Q(x_1, x_2)]Q(x_1, x_2) = P[Q(x_1, x_2)]\left(\frac{\partial Q}{\partial x_1}x_1 + \frac{\partial Q}{\partial x_2}x_2\right)$$

[4] See Allen (1938) pp. 315–322 for a concise treatment of linear homogeneity.

which yields the desired result.

$$P[Q(x_1, x_2)]Q(x_1, x_2) = P[Q(x_1, x_2)]Q(x_1, x_2)$$

to Euler's theorem. This completes the proof of Theorem 4.2.

Theorems 4.1 and 4.2 establish an economic equivalence between vertical integration and tying under input monopoly and variable proportions. In effect, optimal use of the tying alternative results in the downstream producers' return to the efficient expansion path that competitively determined prices would generate. In our model, this requires that the monopolist adjust the price of all tied inputs such that the relative prices remain the same as competitively determined relative prices (i.e., all input prices must be adjusted above marginal cost in equal proportion). At the same time, input prices must exceed marginal costs by an amount that is just sufficient to increase the downstream producers' average cost to the monopoly price in the final product market. This is shown in Figure 4.2 where D is the demand for the final good and MR is the associated marginal revenue. The horizontal curve, MC_1, represents the constant marginal (and average) cost curve for the final good when all inputs are priced at their marginal cost. The maximum profit that this industry can generate requires a price and output of P_M and Q_M, respectively. Under the tying arrangement, the input monopolist will set the input prices such that the

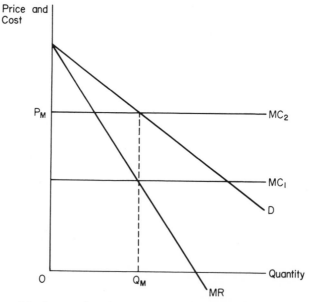

Figure 4.2   Impact of a tying arrangement on the downstream industry.

final-good industry marginal (and average) cost curve shifts from MC_1 to MC_2. Competitive behavior of the downstream firms will lead to the production of Q_M and a price of P_M. Thus, the input monopolist extracts the full profit through the tying arrangement. More formally, these statements can be shown to be true by proof of the following two corollaries that result from Theorems 4.1 and 4.2.

Corollary A The optimal strategy under the tying alternative leads to input prices that exceed marginal costs in equal proportions, that is, $p_1/c_1 = p_2/c_1 > 1$.

Proof Comparison of the downstream industry's first-order conditions, (4.17) and (4.18), with the upstream monopolist's first-order conditions, (4.21) and (4.22), implies

$$p_1 > c_1 \quad \text{and} \quad p_2 > c_2 \tag{4.24}$$

since $(\partial P/\partial Q)(\partial Q/\partial x_1)x_1 + (\partial P/\partial Q)(\partial Q/\partial x_2)x_2 = (\partial P/\partial Q)\, Q < 0$. Dividing (4.17) by (4.18) and (4.21) by (4.22), we obtain

$$\frac{p_1}{p_2} = \frac{\partial Q/\partial x_1}{\partial Q/\partial x_2} = \frac{c_1}{c_2} \tag{4.25}$$

Therefore, (4.24) and (4.25) yield

$$\frac{p_1}{c_1} = \frac{p_2}{c_2} > 1$$

which completes the proof of Corollary A.

Corollary B The optimal strategy under the tying alternative leads to exhaustion of downstream industry profits.

Proof Substituting (4.17) and (4.18) into (4.13), we obtain

$$\pi_T^D = PQ(x_1, x_2) - P\frac{\partial Q}{\partial x_1}x_1 - P\frac{\partial Q}{\partial x_2}x_2$$

Factoring P from the last two terms and applying Euler's theorem,

$$\pi_T^D = PQ(x_1, x_2) - PQ(x_1, x_2) = 0$$

which completes the proof.

Thus, optimal use of the tying alternative results in the existence of a competitively structured downstream industry that combines inputs in efficient proportions (i.e., minimizes costs for given output levels and input prices) and earns zero economic profit. Superficially, there appears to be ideal economic performance, but the industry actually produces and sells the monopoly level of final output, with the implications of this result for the structure, conduct, and performance paradigm of industrial organization economics obvious. The price–quantity constellation that results is completely equivalent to a successful monopolization of the downstream industry by the intermediate-good monopolist through vertical integration. We now turn to our second contractual alternative to ownership integration: output royalties.

Output Royalties

The second contractual alternative to ownership integration involves a per-unit royalty on output. The intermediate-product monopolist may obtain, through a contractual arrangement with downstream firms, the right to place a per-unit tax on the final product.[5] Under such an arrangement, the upstream monopolist's profit function is given by

$$\pi_0^U = p_1(x_1)x_1 + tQ(x_1, x_2, t) - c_1x_1 \qquad (4.26)$$

where t is the unit tax imposed.[6] Behavior at this stage of production is then characterized by maximization of π_0^U over x_1 and t with c_1, $p_1(x_1)$, and $Q(x_1, x_2, t)$ taken as given. Under this strategy, the profit function for the downstream industry is

$$\pi_0^D = PQ(x_1, x_2, t) - p_1x_1 - p_2x_2 - tQ(x_1, x_2, t) \qquad (4.27)$$

Firms in this industry will accept p_1, p_2, P, t, and $Q(x_1, x_2, t)$ as given and maximize π_0^D over x_1 and x_2.

The Equivalence Between Vertical Integration and
Output Royalties

Under the assumptions of the variable proportions model, the input monopolist can achieve the same results as those available through vertical integration or tying if it is possible to obtain the contractual right to place a per-unit royalty on the final output of downstream producers. This assertion can be shown by proving the following two theorems.

Theorem 4.3   Given the assumptions of the variable proportions model, a successful monopolization of the downstream industry

---

[5] For an interesting analysis of ad valorem output royalties imposed by labor unions, see Warren-Boulton (1977).

[6] Industry output Q must be functionally related to the unit tax t in order to reflect a negatively sloped final-product demand. Increasing t raises downstream producers' costs, which, under the assumed condition of a competitive market at the final-good stage, must raise equilibrium output price and lower quantity.

through vertical integration *and* a taxing arrangement whereby the x_1 monopolist sets a per-unit tax on Q both result in inputs x_1 and x_2 being combined in efficient proportions.

Proof   Efficiency of input proportions was established in Theorem 4.1 for the complete vertical integration alternative. To establish the equivalent result under the output taxation alternative, we note that profit maximization on the part of the downstream firms [maximization of π_O^D from expression (4.27)] results in the first-order conditions

$$p_1 = (P - t)\partial Q/\partial x_1 \tag{4.28}$$

and

$$p_2 = (P - t)\partial Q/\partial x_2 \tag{4.29}$$

Solving (4.29) for t and substituting the resulting expression and (4.28) for t and p_1 in (4.26), we obtain the upstream monopolist's profit function which incorporates the optimizing behavior of the firms in the downstream industry:

$$\pi_O^U = P \frac{\partial Q}{\partial x_1} x_1 - \left[\frac{P\,\partial Q/\partial x_2 - p_2}{\partial Q/\partial x_2}\right] \frac{\partial Q}{\partial x_1} x_1 + \left[\frac{P\,\partial Q/\partial x_2 - p_2}{\partial Q/\partial x_2}\right]$$

$$\times\ Q(x_1, x_2, t) - c_1 x_1 \tag{4.30}$$

The intermediate product monopolist maximizes this expression over x_1 and t. Differentiating (4.30) with respect to x_1 and setting the resulting expression equal to 0, we obtain the first necessary condition

$$P \frac{\partial Q}{\partial x_1} + P \frac{\partial^2 Q}{\partial x_1^2} x_1 - \left(\frac{P\,\partial Q/\partial x_2 - p_2}{\partial Q/\partial x_2}\right) \frac{\partial Q}{\partial x_1}$$

$$- \left(\frac{P\,\partial Q/\partial x_2 - p_2}{\partial Q/\partial x_2}\right) \frac{\partial^2 Q}{\partial x_1^2} x_1 - \left[\frac{p_2\,\partial^2 Q/\partial x_2\partial x_1}{(\partial Q/\partial x_2)^2}\right] \frac{\partial Q}{\partial x_1} x_1$$

$$+ \left(\frac{P\,\partial Q/\partial x_2 - p_2}{\partial Q/\partial x_2}\right) \frac{\partial Q}{\partial x_1} + \left[\frac{p_2\,\partial^2 Q/\partial x_2\partial x_1}{(\partial Q/\partial x_2)^2}\right] Q - c_1 = 0 \tag{4.31}$$

The third and sixth terms cancel. Apply Euler's theorem to Q in the seventh term and move c_1 to the right-hand side. Then (4.31) becomes

$$P \frac{\partial Q}{\partial x_1} + P \frac{\partial^2 Q}{\partial x_1^2} x_1 - \left(\frac{P\,\partial Q/\partial x_2 - p_2}{\partial Q/\partial x_2}\right) \frac{\partial^2 Q}{\partial x_1^2} x_1 - \left[\frac{p_2\,\partial^2 Q/\partial x_2\partial x_1}{(\partial Q/\partial x_2)^2}\right] \frac{\partial Q}{\partial x_1} x_1$$

$$+ \left[\frac{p_2\,\partial^2 Q/\partial x_2\partial x_1}{(\partial Q/\partial x_2)^2}\right] \frac{\partial Q}{\partial x_1} x_1 + \left[\frac{p_2\,\partial^2 Q/\partial x_2\partial x_1}{(\partial Q/\partial x_2)^2}\right] \frac{\partial Q}{\partial x_2} x_2 = c_1 \tag{4.32}$$

Now the fourth and fifth terms cancel. Also, upon cancellation, the denominator of the sixth term reduces to $\partial Q/\partial x_2$. Making these cancellations and simplifying the bracketed expression in the third term, (4.32) becomes

$$
P\frac{\partial Q}{\partial x_1} + P\frac{\partial^2 Q}{\partial x_1^2}x_1 - P\frac{\partial^2 Q}{\partial x_1^2}x_1 + \frac{p_2}{\partial Q/\partial x_2}\frac{\partial^2 Q}{\partial x_1^2}x_1
$$

$$
+ \left(\frac{p_2\,\partial^2 Q/\partial x_2\partial x_1}{\partial Q/\partial x_2}\right)x_2 = c_1 \tag{4.33}
$$

The second and third terms cancel. A property of a linearly homogeneous function is that $\partial^2 Q/\partial x_2\partial x_1 = -\partial^2 Q/\partial x_1^2(x_1/x_2)$. Applying this to the last term on the left-hand side, the fourth and fifth terms cancel and the first necessary condition for the maximization of π_O^U becomes

$$
P\,\partial Q/\partial x_1 = c_1 \tag{4.34}
$$

The second necessary condition is obtained by differentiating (4.30) with respect to t and setting the resulting expression equal to 0:

$$
\left(\frac{P\,\partial Q/\partial x_2 - p_2}{\partial Q/\partial x_2}\right)\frac{\partial Q}{\partial t} = 0 \tag{4.35}
$$

Dividing both sides of (4.35) by $\partial Q/\partial t$, multiplying both sides by $\partial Q/\partial x_2$, and recalling that $p_2 = c_2$ under the taxing alternative (because we have asssumed the x_2 market to be competitive), this condition immediately reduces to

$$
P\,\partial Q/\partial x_2 = c_2 \tag{4.36}
$$

Dividing (4.34) by (4.36), we have

$$
\frac{\partial Q/\partial x_1}{\partial Q/\partial x_2} = \frac{c_1}{c_2} \tag{4.37}
$$

and the proof is complete.

Theorem 4.4   Given the assumptions of the variable proportions model, a successful monopolization of the downstream industry through vertical integration *and* a taxing arrangement whereby the x_1 monopolist sets a per-unit tax on Q yield identical profits to the input monopolist.

Proof   We want to show that $\pi_O^U = \pi_I$. Cancelling $c_1 x_1$ from (4.26) and (4.11), this requires that

$$
p_1(x_1)x_1 + tQ(x_1, x_2, t) = P[Q(x_1, x_2)]Q(x_1, x_2) - c_2 x_2 \tag{4.38}
$$

Substituting for p_1 from (4.28) and shortening our notation, (4.38) becomes

$$P \frac{\partial Q}{\partial x_1} x_1 - t \frac{\partial Q}{\partial x_1} x_1 + tQ = PQ - c_2 x_2 \tag{4.39}$$

Applying Euler's theorem to both sides of (4.39) and cancelling, we obtain

$$t \frac{\partial Q}{\partial x_2} x_2 = P \frac{\partial Q}{\partial x_2} x_2 - c_2 x_2 \tag{4.40}$$

Equation (4.29) implies $P(\partial Q/\partial x_2) = p_2 + t(\partial Q/\partial x_2)$. Substituting this on the right-hand side of (4.40),

$$t \frac{\partial Q}{\partial x_2} x_2 = p_2 x_2 + t \frac{\partial Q}{\partial x_2} x_2 - c_2 x_2 \tag{4.41}$$

Therefore,

$$t \frac{\partial Q}{\partial x_2} x_2 = t \frac{\partial Q}{\partial x_2} x_2 \tag{4.41}$$

since $p_2 = c_2$ by assumption. This completes the proof of Theorem 4.4.

Theorems 4.3 and 4.4 establish an economic equivalence between vertical integration and output taxation under input monopoly and variable proportions. Here, however, downstream producers are returned to the efficient expansion path by pricing the monopolized input at marginal cost. Final output producers' average costs are then raised to the monopoly price in the final-good market by adjusting the output royalty rate to equal the difference between the monopoly price and average production costs net of the royalty. Such a taxing scheme will, then, lead to the exhaustion of downstream industry profits. This can be seen by proof of the following three corollaries that result from Theorems 4.3 and 4.4.

Corollary C   Under the output taxation alternative, the intermediate-product monopolist will price x_1 at marginal cost.

Proof   Dividing (4.28) by (4.29),

$$\frac{p_1}{p_2} = \frac{\partial Q/\partial x_1}{\partial Q/\partial x_2}$$

Using (4.37), then

$$\frac{p_1}{p_2} = \frac{c_1}{c_2}$$

which yields the desired result that

$$p_1 = c_1$$

since $p_2 = c_2$.

Corollary D   Under the output taxation alternative, the input monopolist will set the per-unit tax equal to the difference between the final-product price and downstream industry average cost of production net of the tax.

Proof   Equation (4.29) implies that

$$t = P - \frac{p_2}{\partial Q / \partial x_2} \tag{4.43}$$

By the definition of marginal cost MC, we may conclude from (4.43) that

$$t = P - \text{MC}$$

But marginal cost is constant and, therefore, equal to average cost, so

$$t = P - \frac{c_1 x_1 + c_2 x_2}{Q} \tag{4.44}$$

and the proof is complete.

Corollary E   The optimal strategy under the output taxation alternative leads to exhaustion of downstream industry profits.

Proof   Substituting (4.44) into (4.27), we have

$$\pi_O^D = PQ - p_1 x_1 - p_2 x_2 - \left(P - \frac{c_1 x_1 + c_2 x_2}{Q} \right) Q$$

which yields the desired result,

$$\pi_O^U = 0$$

since $p_1 = c_1$ by Corollary C and $p_2 = c_2$ by assumption.

In Figure 4.3, the per-unit output tax is equal to the vertical distance between P_M and MC. Thus, the royalty could be interpreted as an addition to cost that would shift the long-run supply function. Alternatively, the royalty can be interpreted as a wedge between the demand price, which is P_M, and the supply price, which is equal to MC. Under this interpretation, the perceived (or net) demand function, which is denoted by d, intersects MC at Q_M.

As with the tying arrangement, the output royalty mechanism for vertical control results in a competitively structured final-good industry that

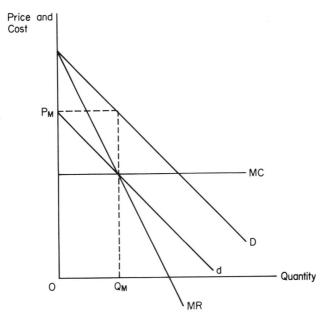

Figure 4.3 Impact of an output royalty on the downstream industry.

combines inputs x_1 and x_2 in efficient proportions and earns zero economic profit yet produces and sells the monopoly level of output Q_M. Again, this result is economically equivalent to a successful monopolization of the downstream industry by the intermediate-good monopolist through vertical integration. Next, we shall consider a close relative of output royalties: sales revenue royalties.

Sales Revenue Royalties

The third contractual alternative to ownership integration involves a royalty on sales revenue. Inaba (1980) provided a proof that this strategy could be used. His is quite different from the following proof. The intermediate-product monopolist may obtain the right to impose an ad valorem tax on sales revenue through a contractual arrangement with the downstream firms. This is a fairly common arrangement between a franchisor and its franchisees. Under such an arrangement, the upstream monopolist's profit function is given by

$$\pi_S^U = p_1(x_1)x_1 + rP[Q(x_1, x_2, r)]Q(x_1, x_2, r) - c_1 x_1 \qquad (4.45)$$

where r is the ad valorem royalty rate. The upstream firm then attempts to maximize π_S^U over x_1 and r with c_1, $p_1(x_1)$, and $Q(x_1, x_2, r)$ taken as given. With a sales revenue royalty, the profit function at the downstream stage is

$$\pi_S^D = (1 - r)PQ(x_1, x_2, r) - p_1 x_1 - c_2 x_2 \qquad (4.46)$$

The downstream firms accept p_1, c_2, P, r, and $Q(x_1, x_2, r)$ as given. Thus, they maximize π_S^D over x_1 and x_2.

The Equivalence between Vertical Integration and Sales Revenue Royalties Under the assumptions of the variable proportions model, the input monopolist can achieve the same results as those available through vertical integration if the contractual right to place an ad valorem royalty on the sales revenue of downstream producers can be obtained. Proof of the following two theorems will demonstrate the truth of this assertion.

Theorem 4.5 Given the assumptions of the variable proportions model, a successful monopolization of the downstream industry through vertical integration *and* a taxing arrangement whereby the x_1 monopolist imposes an ad valorem tax on downstream sales revenue both result in inputs x_1 and x_2 being combined in efficient proportions.

Proof The efficiency of input proportions for the complete ownership integration alternative was established in Theorem 4.1. To establish the equivalent result under the sales revenue royalty alternative, we note that profit maximization on the part of the downstream firms [maximization of π_S^D from expression (4.46)] results in the first-order conditions

$$(1 - r)P \frac{\partial Q}{\partial x_1} - p_1 = 0 \qquad (4.47)$$

and

$$(1 - r)P \frac{\partial Q}{\partial x_2} - c_2 = 0 \qquad (4.48)$$

Solving (4.47) for p_1 and (4.48) for r, we have

$$p_1 = (1 - r)P \frac{\partial Q}{\partial x_1} \qquad (4.49)$$

and

$$r = 1 - \frac{c_2}{P \partial Q / \partial x_2} \qquad (4.50)$$

Substituting (4.50) into (4.49) we obtain

$$p_1 = \frac{c_2\, \partial Q/\partial x_1}{\partial Q/\partial x_2} \tag{4.51}$$

By substituting (4.50) and (4.51) into (4.45) we obtain the upstream monopolist's profit function, taking into account the maximizing behavior of the downstream firms:

$$\pi_S^U = \frac{c_2(x_1\partial Q/\partial x_1 - Q)}{\partial Q/\partial x_2} + PQ - c_1 x_1 \tag{4.52}$$

This function is maximized over x_1 and r. Taking the partial derivative of π_S^U with respect to x_1, we have

$$\frac{\partial \pi_S^U}{\partial x_1} = \frac{c_2\left(x_1 \dfrac{\partial^2 Q}{\partial x_1^2}\dfrac{\partial Q}{\partial x_2} - x_1 \dfrac{\partial Q}{\partial x_1}\dfrac{\partial^2 Q}{\partial x_2 \partial x_1} + Q\dfrac{\partial^2 Q}{\partial x_2 \partial x_1}\right)}{(\partial Q/\partial x_2)^2}$$

$$+ P\frac{\partial Q}{\partial x_1} - c_1 = 0 \tag{4.53}$$

A property of linearly homogeneous functions is that

$$x_1 \frac{\partial^2 Q}{\partial x_1^2} = -x_2 \frac{\partial^2 Q}{\partial x_2 \partial x_1}$$

Substituting this property into (4.53) provides

$$\frac{c_2\left(-x_2 \dfrac{\partial Q}{\partial x_2}\dfrac{\partial^2 Q}{\partial x_2 \partial x_1} - x_1 \dfrac{\partial Q}{\partial x_1}\dfrac{\partial^2 Q}{\partial x_2 \partial x_1} + Q\dfrac{\partial^2 Q}{\partial x_2 \partial x_1}\right)}{(\partial Q/\partial x_2)^2}$$

$$+ P\frac{\partial Q}{\partial x_1} - c_1 = 0 \tag{4.54}$$

Applying Euler's equation, we have

$$\frac{c_2\left(-Q \dfrac{\partial^2 Q}{\partial x_2 \partial x_1} + Q\dfrac{\partial^2 Q}{\partial x_2 \partial x_1}\right)}{(\partial Q/\partial x_2)^2} + P\frac{\partial Q}{\partial x_1} - c_1 = 0$$

or

$$P\, \partial Q/\partial x_1 = c_1 \tag{4.55}$$

Now, partially differentiate (4.52) with respect to r, recalling that $Q = Q(x_1, x_2, r)$:

$$\frac{\partial \pi_S^U}{\partial r} = -\frac{c_2}{\partial Q/\partial x_2} \frac{\partial Q}{\partial r} + P \frac{\partial Q}{\partial r} = 0 \qquad (4.56)$$

Divide both sides of (4.56) by $\partial Q/\partial r$ and multiply by $\partial Q/\partial x_2$:

$$-c_2 + P(\partial Q/\partial x_2) = 0$$

or

$$P(\partial Q/\partial x_2) = c_2 \qquad (4.57)$$

Divide (4.55) by (4.57):

$$\frac{\partial Q/\partial x_1}{\partial Q/\partial x_2} = \frac{c_1}{c_2}$$

and the proof is complete.

Theorem 4.6   Given the assumptions of the variable proportions model, a successful monopolization of the downstream industry through vertical integration *and* a taxing arrangement where by the x_1 monopolist imposes an ad valorem tax on downstream sales revenue both yield identical profits to the input monopolist.

Proof   We want to show that $\pi_S^U = \pi_I$:

$$p_1 x_1 + rPQ - c_1 x_1 = PQ - c_1 x_1 - c_2 x_2 \qquad (4.58)$$

Cancel $c_1 x_1$ from both sides of (4.58) and substitute (4.49) on the left-hand side:

$$(1 - r)P \frac{\partial Q}{\partial x_1} x_1 + rPQ = PQ - c_2 x_2$$

Now, from Euler's theorem, we have $Q = x_1 \partial Q/\partial x_1 + x_2 \partial Q/\partial x_2$. Substituting this for Q and cancelling several terms, we have

$$rP \frac{\partial Q}{\partial x_2} x_2 = rP \frac{\partial Q}{\partial x_2} x_2$$

which completes the proof.

   Theorems 4.5 and 4.6 establish the economic equivalence between vertical integration and sales revenue royalties under input monopoly and variable proportions. Here again the downstream producers are returned to the socially efficient expansion path by pricing the monopolized input at marginal cost. The average costs of the final-good producers are then raised to the monopoly price in the final-good market by adjusting the sales revenue royalty to equal the difference between the monopoly price and the average production costs net of the royalty. Imposition of the optimal sales revenue royalty will lead to the exhaustion of downstream industry profits. This can be seen by proving the following corollaries that result from Theorems 4.5 and 4.6.

Corollary F   Under the sales revenue royalty alternative, the intermediate product monopolist will price x_1 at marginal cost.

Proof   Dividing (4.47) by (4.48) provides

$$\frac{\partial Q / \partial x_1}{\partial Q / \partial x_2} = \frac{p_1}{c_2}$$

But from Theorem 4.5, we know that

$$\frac{\partial Q / \partial x_1}{\partial Q / \partial x_2} = \frac{c_1}{c_2}$$

Thus, the input monopolist must select p_1 such that

$$p_1 = c_1$$

which is the desired result.

Corollary G   Under the sales revenue royalty alternative, the input monopolist will set the royalty rate equal to the percentage difference between the final-product price and downstream industry average cost of production net of the royalty.

Proof   Solving (4.48) for the per-unit royalty rP, we have

$$rP = P - \frac{c_2}{\partial Q / \partial x_2}$$

By the definition of marginal cost MC, we may conclude that

$$rP = P - MC$$

But marginal cost is constant and, therefore, equal to average cost, so

$$r = \left(P - \frac{c_1 x_1 + c_2 x_2}{Q} \right) \bigg/ P \tag{4.59}$$

and the proof is complete.

Corollary H   Selection of the optimal sales revenue royalty rate leads to exhaustion of the downstream profits.

Proof Substituting (4.59) into (4.46), we have

$$\pi_S^D = PQ - p_1 x_1 - c_2 x_2 - \left(P - \frac{c_1 x_1 + c_2 x_2}{Q}\right) Q$$

which yields the desired result:

$$\pi_S^D = 0$$

since $p_1 = c_1$ by Corollary F.

In Figure 4.4, the ad valorem sales revenue royalty causes the perceived demand curve to pass through the point at which marginal revenue MR intersects marginal cost MC. Thus, the royalty should be interpreted as a wedge between the demand price, which is P_M, and the supply price, which is equal to MC.

Once again, the sales revenue royalty alternative for vertical control results in a competitively structured final-good industry that combines inputs x_1 and x_2 in socially efficient proportions and earns 0 economic profit. In spite of this, the downstream industry produces and sells the monopoly level of output Q_M. This contractual result is economically equivalent to a successful monopolization of the downstream industry by the intermediate-good monopolist through vertical integration.

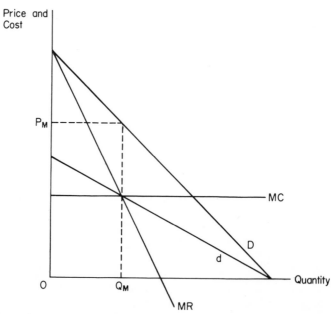

Figure 4.4 Impact of a sales revenue royalty on the downstream industry.

Lump-Sum Entry Fees

A fourth contractual alternative to ownership integration is for the upstream firm to levy a lump-sum fee on downstream producers in exchange for the opportunity to purchase the monopolized input. The possibility of achieving equivalent results through use of a lump-sum entry fee was originally suggested by Philip Coelho. In contrast to the other contractual alternatives, use of this strategy results in the attainment of a monopoly position by one downstream firm (or by n downstream firms where n separate regional markets for final output exist). Following this strategy, then, the intermediate-product monopolist's profit function can be expressed as

$$\pi_F^U = p(x_1)x_1 + F - c_1x_1 \tag{4.60}$$

where F is the lump-sum fee imposed on downstream firms. The monopolist's behavior is then described by maximization of π_F^U over x_1 and F with c_1 and $p_1(x_1)$ taken as given. The profit function of the remaining downstream firm (or firms) will be given by

$$\pi_F^D = P[Q(x_1, x_2)]Q(x_1, x_2) - p_1x_1 - p_2x_2 - F \tag{4.61}$$

Firms at this stage will accept $P[Q(x_1, x_2)]$, p_1, p_2, and F as given and maximize π_F^D over x_1 and x_2.

The Equivalence between Vertical Integration and a Lump-Sum Entry Fee

The final vertical control mechanism described is the imposition of a lump-sum fee on the downstream producers in exchange for the opportunity to purchase the monopolized input x_1. The payment of such a fixed fee by downstream firms will, under the assumed linear homogeneity of $Q(x_1, x_2)$, result in a declining average total cost curve for the final stage of production. This average cost curve will, in fact, be a rectangular hyperbola with marginal production cost (which remains constant) as the horizontal axis. This average total cost curve will be a rectangular hyperbola of the form $ATC = F/Q + MC$ where MC is the constant marginal production cost. Competition among the downstream firms is incompatible with the declining average total costs. Consequently, firms will exit the final stage until there is a single downstream producer. This remaining firm will behave as a monopolist in determining the optimal quantity of final output to produce (i.e., the firm will recognize the effect of output on market price). Despite the monopoly position obtained by this firm, however, it has no monopsony power because the upstream monopolist controls entry into the final stage of production through its control over the input x_1.

The lump-sum fee, as a fixed cost, should not affect the remaining downstream firm's decisions concerning either input proportions or output quantity. It will, however, provide the intermediate-good monopolist an alternative vertical control mechanism that may be used to achieve results that are equivalent to monopolization of the market for Q through vertical ownership integration. This equivalence is established through proof of the following two theorems.

Theorem 4.7   Given the assumptions of the variable proportions model, a successful monopolization of the downstream industry through vertical integration *and* the imposition of a lump-sum entry fee on downstream producers both result in inputs x_1 and x_2 being combined in efficient proportions.

Proof   Efficiency of input proportions was established in Theorem 4.1 for the complete vertical integration alternative. To establish the equivalent result under the entry-fee alternative, we first note that profit maximization on the part of the remaining downstream firm [maximization of π_F^D from expression (4.61)] results in the first-order conditions

$$p_1 = \left(P + Q \frac{\partial P}{\partial Q} \right) \frac{\partial Q}{\partial x_1} \tag{4.62}$$

or

$$p_2 = \left(P + Q \frac{\partial P}{\partial Q} \right) \frac{\partial Q}{\partial x_2} \tag{4.63}$$

Solving (4.63) for P and substituting the resulting expression in (4.62), we obtain

$$p_1 = \frac{p_2}{\partial Q / \partial x_2} \frac{\partial Q}{\partial x_1} \tag{4.64}$$

Substituting (4.64) into π_F^U from expression (4.60), we obtain the upstream monopolist's profit function which incorporates the optimizing behavior of the remaining firm at the downstream stage:

$$\pi_F^U = \frac{p_2}{\partial Q / \partial x_2} \frac{\partial Q}{\partial x_1} x_1 + F - c_1 x_1 \tag{4.65}$$

The upstream firm attempts to maximize π_F^U over x_1 and F.

Such maximization, however, will be subject to the constraint that F cannot exceed the downstream firm's total profits net of the entry fee. Otherwise, the downstream firm will not be able to remain in business. Since the final-good producer's marginal and average production costs net of the fee are constant at $p_2/(\partial Q/\partial x_2)$, the upstream monopolist's optimization problem can then be stated as

$$\max_{x_1, F} \pi_F^U$$

subject to

$$F \le \left(P - \frac{p_2}{\partial Q/\partial x_2}\right) Q$$

Forming the Lagrangian,

$$L = \frac{p_2}{\partial Q/\partial x_2} \frac{\partial Q}{\partial x_1} x_1 + F - c_1 x_1 + \lambda\left[\left(P - \frac{p_2}{\partial Q/\partial x_2}\right) Q - F\right] \quad (4.66)$$

where λ is the Lagrange multiplier, the Kuhn-Tucker first-order conditions are:

$$\frac{\partial L}{\partial x_1} = \frac{p_2}{\partial Q/\partial x_2} \frac{\partial Q}{\partial x_1} + \frac{p_2}{\partial Q/\partial x_2} \frac{\partial^2 Q}{\partial x_1^2} x_1 - \frac{p_2}{(\partial Q/\partial x_2)^2} \frac{\partial^2 Q}{\partial x_1 \partial x_2} \frac{\partial Q}{\partial x_2} x_1 - c_1$$

$$+ \lambda\left[P \frac{\partial Q}{\partial x_1} - \frac{p_2}{\partial Q/\partial x_2} \frac{\partial Q}{\partial x_1} + Q \frac{\partial P}{\partial Q} \frac{\partial Q}{\partial x_1} + \frac{p_2}{(\partial Q/\partial x_2)^2} \frac{\partial^2 Q}{\partial x_1 \partial x_2} Q\right]$$

$$\le 0 \quad (4.67)$$

$$\frac{\partial L}{\partial x_1} x_1 = \left\{\frac{p_2}{\partial Q/\partial x_2} \frac{\partial Q}{\partial x_1} + \frac{p_2}{\partial Q/\partial x_2} \frac{\partial^2 Q}{\partial x_1^2} x_1 - \frac{p_2}{(\partial Q/\partial x_2)^2} \frac{\partial^2 Q}{\partial x_1 \partial x_2} \frac{\partial Q}{\partial x_2} x_1 - c_1\right.$$

$$+ \lambda\left[P \frac{\partial Q}{\partial x_1} - \frac{p_2}{\partial Q/\partial x_2} \frac{\partial Q}{\partial x_1} + Q \frac{\partial P}{\partial Q} \frac{\partial Q}{\partial x_1}\right.$$

$$\left.\left.+ \frac{p_2}{(\partial Q/\partial x_2)^2} \frac{\partial^2 Q}{\partial x_1 \partial x_2} Q\right]\right\} x_1 = 0 \quad (4.68)$$

$$x_1 \ge 0 \quad (4.69)$$

$$\frac{\partial L}{\partial F} = 1 - \lambda \le 0 \quad (4.70)$$

$$\frac{\partial L}{\partial F} F = (1 - \lambda) F = 0 \quad (4.71)$$

$$F \ge 0 \quad (4.72)$$

$$\frac{\partial L}{\partial \lambda} = \left(P - \frac{p_2}{\partial Q/\partial x_2}\right) Q - F \ge 0 \quad (4.73)$$

$$\frac{\partial L}{\partial \lambda} \lambda = \lambda\left[\left(P - \frac{p_2}{\partial Q/\partial x_2}\right) Q - F\right] = 0 \quad (4.74)$$

and

$$\lambda \geq 0 \tag{4.75}$$

If the downstream firm produces a positive level of final output, $x_1 > 0$, and (4.67) will hold as an equality. Then, if the intermediate-product monopolist imposes a positive entry fee, $F > 0$, and (4.70) will hold as an equality. From (4.70), then, $\lambda = 1$. Substituting this into (4.67), we have

$$\frac{p_2}{\partial Q/\partial x_2} \frac{\partial Q}{\partial x_1} + \frac{p_2}{\partial Q/\partial x_2} \frac{\partial^2 Q}{\partial x_1^2} x_1 - \frac{p_2}{(\partial Q/\partial x_2)^2} \frac{\partial^2 Q}{\partial x_1 \partial x_2} \frac{\partial Q}{\partial x_1} x_1 + P \frac{\partial Q}{\partial x_1}$$

$$- \frac{p_2}{\partial Q/\partial x_2} \frac{\partial Q}{\partial x_1} + Q \frac{\partial P}{\partial Q} \frac{\partial Q}{\partial x_1} + \frac{p_2}{(\partial Q/\partial x_2)^2} \frac{\partial^2 Q}{\partial x_1 \partial x_2} Q = c_1 \tag{4.76}$$

The first and fifth terms cancel. Linear homogeneity implies $(\partial^2 Q/\partial x_1^2)x_1 = -(\partial^2 Q/\partial x_1 \partial x_2)x_2$. Substituting this into the second term on the left-hand side and multiplying this term by $(\partial Q/\partial x_2/(\partial Q/\partial x_2))$, expression (4.76) becomes

$$- \frac{p_2}{(\partial Q/\partial x_2)^2} \frac{\partial^2 Q}{\partial x_1 \partial x_2} \frac{\partial Q}{\partial x_2} x_2 - \frac{p_2}{(\partial Q/\partial x_2)^2} \frac{\partial^2 Q}{\partial x_1 \partial x_2} \frac{\partial Q}{\partial x_1} x_1 + P \frac{\partial Q}{\partial x_1}$$

$$+ Q \frac{\partial P}{\partial Q} \frac{\partial Q}{\partial x_1} + \frac{p_2}{(\partial Q/\partial x_2)^2} \frac{\partial^2 Q}{\partial x_1 \partial x_2} Q = c_1 \tag{4.77}$$

Factoring $-[p_2/(\partial Q/\partial x_2)^2] (\partial^2 Q/\partial x_1 \partial x_2)$ from the first two terms and applying Euler's theorem, the first, second, and fifth terms cancel, leaving

$$\left(P + Q \frac{\partial P}{\partial Q}\right) \frac{\partial Q}{\partial x_1} = c_1 \tag{4.78}$$

Dividing (4.78) by (4.63) and noting that $p_2 = c_2$ by assumption, we have

$$\frac{\partial Q/\partial x_1}{\partial Q/\partial x_2} = \frac{c_1}{c_2} \tag{4.79}$$

which completes the proof of Theorem 4.7.

Theorem 4.8   Given the assumptions of the variable proportions model, a successful monopolization of the downstream industry through vertical integration *and* the imposition of a lump-sum entry fee on downstream producers yield identical profits to the input monopolist.

Proof   We want to show that $\pi_F^U = \pi_I$. Cancelling $c_1 x_1$ from (4.11) and (4.60), this requires that

$$p_1(x_1)x_1 + F = P[Q(x_1, x_2)]Q(x_1, x_2) - c_2 x_2 \tag{4.80}$$

With x_1, $F > 0$, (4.71) implies $\lambda = 1$. This means that (4.73) will be met as an equality and, consequently, that

$$F = \left(P - \frac{p_2}{\partial Q/\partial x_2}\right)Q \qquad (4.81)$$

Substituting (4.64) for p_1 and (4.81) for F, (4.80) becomes

$$\frac{p_2}{\partial Q/\partial x_2}\frac{\partial Q}{\partial x_1}x_1 + PQ - \frac{p_2}{\partial Q/\partial x_2}Q = PQ - c_2 x_2$$

Applying Euler's theorem to Q in the third term on the left-hand side, we have

$$\frac{p_2}{\partial Q/\partial x_2}\frac{\partial Q}{\partial x_1}x_1 + PQ - \frac{p_2}{\partial Q/\partial x_2}\frac{\partial Q}{\partial x_1}x_1 - \frac{p_2}{\partial Q/\partial x_2}\frac{\partial Q}{\partial x_2}x_2 = PQ - c_2 x_2$$

The first and third terms cancel, and the fourth term simplifies, leaving

$$PQ - c_2 x_2 = PQ - c_2 x_2,$$

since $p_2 = c_2$ by assumption. This completes our proof of Theorem 4.8.

Unlike the output tax or the tying arrangement, the entry-fee alternative will result in a monopoly structure at the final output stage. As under the output taxation alternative, however, optimal use of the lump-sum fee leads the upstream firm to price the monopolized input at marginal cost. The fixed fee will then be used to increase the downstream monopolist's average costs to equality with the final-product price, thereby transferring full monopoly profits to the upstream firm. This may be seen by proof of the following corollaries.

Corollary I Under the entry-fee alternative, the input monopolist will price x_1 at marginal cost.

Proof Dividing (4.62) by (4.63),

$$\frac{p_1}{p_2} = \frac{\partial Q/\partial x_1}{\partial Q/\partial x_2}$$

Using (4.79),

$$\frac{p_1}{p_2} = \frac{c_1}{c_2}$$

which yields the desired result that

$$p_1 = c_1$$

since $p_2 = c_2$.

Corollary J Under the entry-fee alternative, the input monopolist will set the lump-sum fee equal to the difference between the total revenue and total production (net of fee) cost of the remaining downstream firm.

Proof This follows immediately from inspection of (4.81) since $p_2/(\partial Q/\partial x_2)$ equals constant marginal and average production costs.

Corollary K The optimal strategy under the entry-fee alternative leads to exhaustion of the downstream firm's profits.

Proof Substituting (4.81) into (4.61),

$$\pi_F^D = PQ - p_1 x_1 - p_2 x_2 - PQ + \frac{p_2}{\partial Q/\partial x_2} Q$$

The first and fourth terms on the right-hand side cancel. Applying Euler's theorem to the final term, we have

$$\pi_F^D = - p_1 x_1 - p_2 x_2 + \frac{p_2}{\partial Q/\partial x_2} \frac{\partial Q}{\partial x_1} x_1 + p_2 x_2$$

Now the second and final terms cancel. Substituting for p_1 from (4.64),

$$\pi_F^D = - \frac{p_2}{\partial Q/\partial x_2} \frac{\partial Q}{\partial x_1} x_1 + \frac{p_2}{\partial Q/\partial x_2} \frac{\partial Q}{\partial x_1} x_1 = 0$$

and the proof is complete.

The operation of the fixed-fee alternative is depicted in Figure 4.5. Given a monopoly position at the final stage of production, the remaining downstream firm will maximize profits by producing Q_M and selling it at P_M. The presence of the fixed fee will raise the average total cost curve above the marginal cost curve, with $(ATC - MC)Q$ equal to the lump-sum fee by definition. Increasing the fee will shift ATC up and to the right while decreasing the fee will have the opposite effect. In the graph, the fixed fee has been adjusted so that $ATC = P_M$ at Q_M. This represents the optimal level of F from the upstream monopolist's point of view. A lower entry fee would permit the downstream firm to capture a share of the monopoly profits, and a higher fee would force the final-good producer out of the market.

If the total output of the final product is sold in n geographically separate markets, then the upstream monopolist may sell the right to produce in each of these markets to a different entrepreneur. The fixed fee charged each prospective downstream monopolist will, then, vary directly with the level of demand in the relevant market. By maintaining the fee precisely at the level given in (4.81) or, equivalently, the shaded area in Figure 4.5 in the various geographic markets, the upstream firm

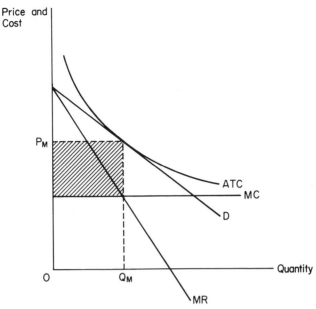

Figure 4.5   Impact of a lump-sum entry fee on the downstream firm.

may capture the full monopoly rents available at the final stage of production.

An important distinction between the lump-sum fee and the other three
vertical control mechanisms arises when production at the final-good
stage is not subject to constant costs. In the presence of increasing marginal (and U-shaped average) costs at the firm level, the fixed-fee alternative will fail to yield equivalent profits to the upstream monopolist. This
may be seen in Figure 4.6. With U-shaped average costs, imposition of a
fixed fee will not result in the "natural monopoly" declining costs drawn
in Figure 4.5 but will, instead, shift the downstream firms' average total
cost curve up and to the right, leaving marginal costs unaffected. The
downstream industry will, then, continue to behave competitively, setting
price equal to marginal cost. Exit will occur as a result of imposition of the
fee, however, for two reasons. First, total industry output will decline
from Q_c to Q_M as the fee raises average total costs from ATC to ATC'.
And second, per-firm output will increase from q_0 to q_1 as remaining firms
move up the marginal cost curve to the minimum point on ATC'. This
latter effect, however, will have the undesirable result that average variable costs will be increased above the minimum point on ATC. Consequently, of the total monopoly profits available in the final-product market

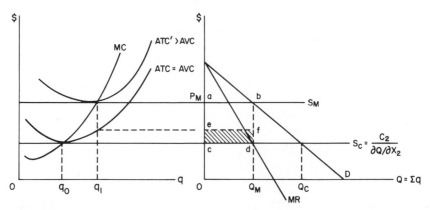

Figure 4.6 Impact of a lump-sum entry fee with U-shaped average costs at the down-stream stage.

(given by the area abcd) some portion (given by the shaded area cdef) will be lost to inefficient production scale in the remaining downstream firms. Furthermore, as a result of this imperfection in the operation of the fixed-fee alternative in the presence of increasing firm costs, the upstream monopolist attempting to employ this strategy may find the profit-maximizing level of the fee to be above that which raises average total costs to ATC'. Thus, the supply curve in the final-good market may be shifted above S_M under this alternative.

None of the other vertical control strategies will be subject to this problem. Under ownership integration, the monopolist will produce q_0 units of final output in each plant and maintain a sufficient number of plants to produce Q_M total units. Under both the tying and the output taxation alternatives, downstream firms' marginal and average costs will be shifted vertically, leaving q_0 the optimal quantity of the final product to manufacture in each plant. Sufficient exit will then occur to yield a total industry output of Q_M. Average variable costs of production, then, will be maintained at the minimum point on AVC under the other three control options.

Theorems 4.1–4.8 establish the economic equivalence of five vertical control mechanisms: vertical integration, tying arrangements, output royalties, sales revenue royalties, and entry fees. The first involves the internalization of the decision-making process of downstream firms by ownership control while the last four involve control exercised through contractual agreements. Together, these five basic strategies describe and provide a foundation for understanding a host of specific business practices that have puzzled economists for some time.

Within the confines of our simplified model, the intermediate-product monopolist should be indifferent as to which of these strategies to employ inasmuch as they all yield equal profits. In practice, however, extraneous factors, primarily in the form of transaction and contracting costs, will operate to influence the choice of which mechanism to apply in any given set of circumstances. The following section briefly describes the kinds of problems likely to be encountered in attempting to implement these alternative forms of vertical control.

Vertical Controls: Problems in Practice

We have seen that when an input monopolist has an incentive to exercise vertical control, there are at least five alternative strategies available. (Some combinations of these alternatives will also work. See Blair and Kaserman's comment on Inaba, 1982.) The monopolist, however, is unlikely to be indifferent when choosing among these control options because the nonproduction costs are not apt to be the same for each alternative. Up to this point, we have ignored most of the costs and complexities associated with implementing the various alternatives in practice. We turn to these considerations here.

Ownership Integration

Vertical control in the form of ownership integration can be expected to represent a superior alternative in situations in which contracting costs are high and significant managerial diseconomies of scale are absent. Some of the factors that may tend to increase the costs of negotiating and enforcing vertical control contracts are examined. (See Williamson, 1971, for a further discussion of these factors.) In addition to the inherent complexities of the managerial function in particular production settings, the potential severity of managerial diseconomies would appear to depend upon the number of plants of efficient scale required for the manufacture of final output. Where efficient producing units are large in relation to total industry output, relatively few plants will be required, and problems of internal interplant control will be likely to be at a minimum. For this reason, vertical control through ownership integration may be relatively attractive in large-scale manufacturing industries as a means of economizing on contracting costs.

Aside from possible decreasing returns to the managerial function, however, two additional factors may tend to discourage the use of this much-investigated control alternative. First, as shown by Vernon Smith

(1971), the cost of capital may be an increasing function of the quantity of capital demanded by the firm. Thus, ownership integration may entail significantly higher capital costs, which could potentially reduce the monopoly profits obtainable through the use of this strategy. Second, some state statutes may render vertical integration illegal. For example, a Maryland statute (Md. Ann. Code, Art. 56, §157C), recently upheld by the U.S. Supreme Court, forbids vertical integration by a producer or refiner of gasoline forward into retailing. [See *Exxon Corporation* v. *Governor of Maryland*, 437 U.S. 117 (1978).] Finally, monopolization of the final-product market, which is necessary for the maximization of profits under the ownership integration alternative, may expose the intermediate-product monopolist to antitrust attack under either Section 2 of the Sherman Act (if integration is carried out through internal expansion) or Section 7 of the Clayton Act (if vertical merger or acquisition is employed). We shall defer an extensive analysis of these matters until Chapter 7.

Under the conditions we have analyzed, the input monopolist can ascertain its unique profit maximizing output. Consequently, as it begins to vertically integrate forward by internal expansion, the upstream monopolist will begin competing with the remaining downstream firms. These firms will eventually have to leave the industry, but only actual or prospective losses will induce exit when there can be no buyer of the firm's goodwill. This will create the aura of predation. Familiar complaints regarding price squeezes (see Singer, 1968, Chapter 19) and predatory refusals to deal will be heard. There is some legal foundation for the argument that the intermediate-product monopolist has an obligation to supply the downstream firms because they cannot survive without the monopolist's product (see Areeda, 1981, pp. 217–220 and 279–280). Consequently, contractual alternatives to ownership integration may provide superior control mechanisms in certain situations. Inspection reveals, however, that these strategies are not without practical problems either. Where the use of such contractual alternatives is (for reasons discussed in the following) infeasible, the upstream monopolist may employ partial vertical integration as a second-best strategy. Schmalensee (1973) shows that $\partial \pi / \partial \lambda > 0$, where λ is the fraction of total industry final output produced by the downstream subsidiaries of the partially integrated monopolist. An optimum $\lambda < 1$ might, then, be found where the expected costs of further increases in λ attributable to antitrust prosecution just offsets the gains to be realized in increased profits.

Tying Arrangements

A tying arrangement may be implemented by fixing input proportions directly by contract (as in many franchising and labor agreements) or by pricing the substitutable inputs appropriately (Corollary A). The advantages that tying may offer over vertical ownership integration are potential savings in capital costs and avoidance of managerial diseconomies of scale.

Several serious difficulties, however, are likely to be encountered in attempting to employ this contractual alternative. First, if the production function for final output admits many inputs that are substitutable for the monopolized input, a tying arrangement involving all such inputs may be prohibitively complex to negotiate and enforce.

Second, even if the tying arrangement involves only one substitutable input (x_2), it will have to be policed. Although by hypothesis the production of x_1 is monopolized, the supply of the other (substitutable) input is not likely to be monopolized. As a result, the downstream firms will have an incentive to buy some x_2 at the competitive (lower) price on the open market. This can be used along with the x_2 purchased from the monopolist. In this way, a limited amount of substitution away from the use of x_1 will still be experienced. Consequently, a requirements contract may have to be imposed on downstream producers in order to circumvent this second-order form of substitution. Such contracts, however, do not entirely remove the policing costs involved in the use of this alternative. If there is some additional profit to be gained from breaching a contract, the terms of the contract will have to be enforced.

Third, and, perhaps more important, use of the tying alternative exposes the upstream monopolist to a greater threat of antitrust attack than is experienced with any of the other vertical control options. Supreme Court precedent suggests that tying arrangements are apt to be found illegal per se. Such precedent indicates that a tying contract is in violation of Section 1 of the Sherman Act when (1) the tying product confers sufficient economic power upon the supplier to control prices and (2) a not-insubstantial dollar volume of commerce in the tied product has been affected. If either condition (1) or condition (2) is satisfied, then Section 3 of the Clayton Act is offended. [The distinction between burdens of proof was specified in *Times-Picayune Publishing Co.* v. *United States,* 345 U.S. 594 (1953).] Experience indicates that it is very easy to satisfy the requirements for per se illegality under this rule. For example, in *Northern Pacific Railway Co.* v. *United States,* 345 U.S. 1 (1958), the Supreme Court found sufficient power in the unique and strategic location of the railroad's land (the tying good). The existence of a copyright on the tying good provided sufficient economic power in *United States* v. *Loew's Inc.,* 371 U.S. 38 (1962). In *Loew's,* a dollar volume of $60,800 was found to be not insubstantial.

It is clear that tying is a very hazardous way of extracting monopoly rents. Violations of the antitrust laws subject the offender to treble-damage suits. Tying allegedly imposes damages upon the monopolist's competitors who are foreclosed from selling the monopolized inputs and upon its customers who are charged extra-competitive prices on the non-monopolized inputs. It is clear from our analysis that tying imposes no injury upon the purchasers of the tied inputs because alternative contractual arrangements permit purchase at competitive prices but do not permit higher profits. This does not necessarily mean that no damages will be awarded to the monopolist's customers. Such a possibility has a chilling effect on economic incentives. Areeda (1976) discussed this problem in the context of a specific case, but did not offer a general argument. More will be said about tying in Chapter 8.

Output and Sales Revenue Royalties

The output and sales revenue royalty alternatives offer several attractive features from the point of view of the upstream monopolist. First, as with the tying alternative, the firm is able to realize capital cost savings and, at the same time, avoid potentially significant managerial diseconomies of scale. Second, unlike the tying option, the output or sales revenue royalty avoids the need for requirements contracts and does not involve the negotiation of complex multiple-input agreements. Third, as was indicated in the preceding section, "Lump-Sum Entry Fees," unlike the entry fee, the royalty agreements can function properly in the presence of U-shaped cost curves at the downstream stage. And finally, these control options maintain the competitive market structure observed at the downstream stage and are relatively safe from antitrust attack.

It appears that the only difficulty likely to be encountered in the implementation of this strategy lies in the need to monitor the output level of all firms at the downstream stage in order to guard against tax evasion. With x_1 priced at marginal cost and the downstream production function known, however, Q can be monitored on the basis of x_1 sales. This will work perfectly, however, only if the downstream firms are unaware of the counting device. Otherwise, there will be an incentive to evade part of the tax by substituting away from x_1 in favor of x_2. Nonetheless, the overall attractiveness of the royalty alternatives may account for their widespread use in labor, patent, and franchising agreements.

Entry Fee

The entry-fee alternative offers many of the same advantages as the output and sales revenue royalties. Managerial diseconomies are avoided, capital costs are kept at a minimum, requirements contracts and complex input agreements are not needed, and the antitrust enforcement authorities are not flagrantly taunted. In addition, the entry fee does not require constant monitoring of final-good production.[7] There are, however, two potentially serious drawbacks to the use of this strategy.

First, as noted in the preceding section, use of this alternative under constant cost production at the downstream stage will lead to a monopolization of the final-product market (or markets). While such monopolization may not openly invite antitrust action when multiple geographic markets are involved, it may arouse some suspicion. Moreover, the presence of U-shaped cost curves at the downstream stage renders the entry fee an inferior alternative for vertical control (Figure 4.6). And second, the entry-fee alternative is likely to encounter substantial problems of an informational nature in practice. Calculation of the optimal level of the fee by the upstream monopolist requires knowledge of the demand curve for final output in each geographic market not only in the present period but in all future periods to which the fee applies. In addition, there will exist serious informational problems on the part of the downstream firm (or firms) under this alternative. Divergent estimates of the future profitability of producing Q over some specified period may render a mutually agreeable entry fee nonexistent. Even under uniform expectations, divergent attitudes toward risk may yield the same result. Moreover, the downstream producer is not apt to pay the present value of full monopoly profit without some guarantee of being able to collect such profit in the future (see Bierman and Tollison, 1970). Such guarantees may be extremely difficult to provide and, in the case of multiple geographic markets, may have to be buttressed with exclusive dealing contracts or territorial restrictions. For these reasons, the entry-fee alternative is, perhaps, not as widely used as is the output tax.

Given these practical difficulties, which are likely to be associated with the various forms of vertical control discussed, the optimal strategy for the input monopolist to employ will, obviously, vary according to the particular set of circumstances encountered. Furthermore, since none of the available control mechanisms is likely to function perfectly, the monopolist must make a selection from a feasible set of second-best alternatives. Consequently, a combination of the various schemes may prove to be preferable in many real-world situations (some franchise agreements observed in practice contain elements of all four control mechanisms dealt with here). (See Blair and Kaserman, 1982, for an analysis of entry fees combined with output royalties in a franchising situation.)

---

[7] As indicated previously, however, calculation of the optimal fee will require knowledge of the demand curve (or curves) at the final-product stage. For a discussion of the use of fixed fees in franchising agreements, see Caves and Murphy (1976).

Conclusion

As noted above, the direction of the social welfare effect of an extension of monopoly power under the conditions postulated in the variable-proportions model has been shown to depend upon the specific production and demand conditions present. Whatever the welfare significance of vertical integration, however, the same significance should be attached to its contractual equivalents. Consequently, the antitrust significance of the economic equivalents ought to be the same. In fact, however, this is not the case at all. The particular form of vertical control matters very much to the judiciary.

A recent summary statement of the judicial concern for the form of vertical control employed can be found in Justice White's concurring opinion in the *GTE Sylvania* decision [*Continental TV, Inc.* v. *GTE Sylvania, Inc.*, 433 U.S. 36 (1977)]. It would not be unfair to characterize White's view as ambivalent. He certainly appreciates the economic approach to analyzing vertical restraints. This is evidenced by his grasp of the fact that there is considerable similarity between nonprice vertical restraints and vertical price fixing. Despite a recognition of the economic efficiency analysis, however, he is unwilling to give up the prohibitions that have applied unequally to both kinds of restraints. To justify this position, he focuses on the preservation of economic freedom for the individual businessman. Although the resulting system may not be the most efficient, those adopting White's view are willing to sacrifice some economic efficiency to maximize the freedom of the independent businessman to conduct his business as he sees fit. But such freedom may be more apparent than real. To the extent that legally acceptable vertical controls can be substituted for what are perceived to be more offensive economic equivalents, efforts at preserving freedom will fail. Nonetheless, as White's view makes some arrangements more dangerous than others, it will bias the input monopolist's choice among the available vertical control alternatives. Such an outcome may tend to encourage the adoption of comparatively safe control mechanisms in situations in which strict efficiency considerations would dictate the use of an otherwise equivalent alternative.

5

Vertical Integration under Uncertainty

The role that uncertainty plays in influencing the decisions of firms to integrate vertically is both diverse and pervasive. We have already discussed the channels through which market uncertainty increases transaction costs and have shown how the presence of such costs creates a profit incentive for vertical integration (Chapter 2). In addition, we shall see that the myriad sources of uncertainty that are typically encountered can create a host of incentives for vertical integration that are independent of traditional transaction cost considerations. In this chapter, we will show how vertical integration can be employed to reduce uncertainty in intermediate-product supply or demand, to assure a more certain source of supply of some essential input or demand for output, to improve the flow of information between vertically related stages of production, to shift risk either to or away from the integrating firm, and to reduce the overall level of risk experienced through diversification.

The major significance of uncertainty in explaining vertical integration has long been recognized. Coase (1937, p. 338) stated, "It seems improbable that a firm would emerge without the existence of uncertainty." Despite this early recognition of the importance of the relationship between uncertainty and vertical integration, however, relatively little work was done on this topic outside the transaction costs literature until very recently. Prior to the mid-1970s, only two studies, both of which were published during the early 1960s, had addressed this subject. Since the mid-1970s, however, this literature has grown rapidly.

We review this body of literature in the present chapter. Following this review, we present some additional analysis of the potential role of random elements in both the demand for final output and the supply of an input in decisions to integrate vertically. At each stage, we indicate the welfare effects of the vertical integration that arises from the particular incentive under investigation. Finally, we briefly discuss some contractual alternatives to vertical integration in the presence of uncertainty.

Early Studies

The first study to examine the relationship between uncertainty and vertical integration in a non-transaction-cost framework is that of Jensen, Kehrberg, and Thomas (1962). This paper is concerned primarily with vertical integration in agricultural product markets, and views vertical integration as a method for reducing the variability of supply and/or demand at vertically related stages of production. Since such variability imposes real resource costs in the form of stockpiles, insurance schemes, and product spoilage, any reduction obtained increases the profits of one or more of the parties to the transaction. Moreover, it is argued that the extent of the variation that is experienced in the absence of vertical integration is directly related to the perishability of the product and the discreteness of the production process involved (i.e., the degree to which the flow of output is lumpy). Consequently, it is predicted that products that are highly perishable and that are characterized by relatively discrete production processes will be likely candidates for vertical integration. While the study provides no detailed explanation of how vertical integration leads to a reduction in the variability of supply or demand for the intermediate product, an argument is made that improved communication and information flows account for this result (p. 379). By focusing on the informational advantages of vertical integration, this study is similar to the earlier work of Malmgren (1961) and the later work of Arrow (1975). By also focusing on the role of vertical integration in reducing variability in the intermediate-product market, it bears some resemblance to the later work of Bernhardt (1977).

The second of these early studies of uncertainty and vertical integration, Wu (1964), is similar in spirit to the first. Following a rigorous treatment of vertical integration under conditions of bilateral monopoly, successive monopoly, and monopsony with no uncertainty, Wu examined the influence of fluctuating input demand on the intermediate firm's profits in the absence of vertical integration. [This was the first formal analysis of vertical integration under monopsony of which we are aware. It remained the only treatment of this subject until the works of McGee and Bassett (1976) and Perry (1978c).] The analysis implicitly assumes risk neutrality by adopting the objective of expected profit maximization. Assuming imperfect forecasting capabilities, a discounted loss function is specified reflecting the opportunity costs that the firm experiences as a consequence of the resulting nonoptimal levels of production, sales, and inventory holdings. These expected losses are then treated as a form of operating cost in the firm's profit function. It is shown that, due to these costs, the firm will produce a lower quantity of output, charge a higher price, and earn less profit under uncertainty than it would under conditions of static demand or perfect forecasting. Moreover, the reduction in output in each period due to conditions of risk is shown to be directly related to the degree of variability of demand around its expected level.

Wu then argued that vertical integration will reduce or eliminate the variation in demand experienced by the upstream producer, and because of this reduction in uncertainty, expected profits will increase. Once again, the precise mechanism through which vertical integration reduces demand variations is left unspecified. Apparently, input demand fluctuations are attributable to some form of transactional uncertainty of the variety later analyzed by Carlton (1979), because Wu indicated that the burden of demand uncertainty is not eliminated from the upstream industry but is merely shifted from the integrated producers to the nonintegrated producers.

More Recent Studies

Following Wu (1964), a full decade passed before any significant new work appeared relating vertical integration to uncertainty. Since 1974, however, several papers have addressed this topic. First, in that year, Green produced a working paper in which vertical integration was undertaken to assure the supply of an input or the demand for output. In Green's (1974) model, the incentive to integrate is created by an assumed price rigidity in the intermediate-product market that leads to production shortages or surpluses that, in turn, lead to rationing on the demand and supply sides of that market, respectively. It is assumed that markets are otherwise competitive, that there exist no regulatory constraints or taxes, and that both the intermediate and final products are obtained from fixed-coefficient production functions.

Rationing occurs on an all-or-nothing basis to the individual firms with the percent of firms so rationed (input producers in situations of excess supply and input consumers in situations of excess demand) varying with fluctuations in exogenous excess demand. In effect, this results in a constant (exogenous) price to unrationed firms participating in the intermediate product market. Rationed firms, on the other hand, are confronted with an effective price of 0 or infinity according to whether they are sellers or buyers in this market. Also, the intermediate good is assumed to be nondurable. As noted by Green (1974, p. 32), the extreme nature of the rationing system employed in this model and the assumed inability of individual firms to affect the probability of being subjected to such rationing (through, e.g., long-term contracts or interfirm ties of a less formal variety) lead, in large measure, to the sharpness of the results obtained.

Since integrated firms are allowed to transfer the output of their upstream divisions directly into the next stage of production without facing the risk of being rationed, an incentive for vertical integration exists in the potential for circumventing an imperfect market. Against this incentive, a countervailing influence is assumed to exist in the form of decreased technological efficiency of larger enterprises (i.e., diseconomies of scale). The firm must then weigh the benefits of avoiding the risk of being rationed against the costs of enlarging its scale of activities through vertical integration.

Two important results emerge from this model. First, under the assumptions employed, the only organizational structure that is optimal for a vertically integrated firm is the balanced one in which the firm does not participate in the intermediate-product market at all. In other words, the firm produces its entire requirements for the given input and transfers its entire production of this input internally. Second, the model is inherently unstable at equilibria in which a mixed organizational structure is present (in the sense that both integrated and nonintegrated firms coexist). The market tends to be propelled toward the extremes of complete or no integration, depending upon the frequency of rationing and the importance of scale diseconomies.

Clearly, if the former occurs and all firms achieve balanced integration, then the intermediate-product market will disappear altogether. Thus, the model not only provides a potential incentive for vertical integration, but it also suggests certain dynamic structural consequences of such integration. The integration of successive stages by one or a few firms may set in motion a chain of events whereby market structure tends to gravitate toward complete and balanced vertical integration. The possibility of important dynamic structural effects is discussed briefly in Edwards (1953), Comanor (1967), and Mueller (1969). The importance of such potential effects, however, is open to debate.

Next, Arrow (1975) examined the influence of stochastic input supply price on the firm's organizational choice when there exists asymmetric information between participants at the upstream and downstream stages. Here, production of the final product is assumed to occur under constant cost conditions with variable proportions and two inputs, capital and raw material. Assuming that downstream firms must purchase their capital one period in advance of production and that upstream firms have information concerning the supply price of the raw material one period in advance (where this price is subject to random variation), the acquisition of raw material suppliers by final-product firms reduces costs at the downstream stage by allowing inputs to be purchased in efficient proportions.

Essentially, vertical integration is seen as a means of acquiring predictive information concerning the relative prices of inputs at the relevant point in time. The need to integrate to acquire this information must, as Arrow pointed out, ultimately rest on the assumption that a market exists for upstream firms but that a market for the information itself does not exist (perhaps for reasons associated with appropriability problems). Also, it must be assumed that a futures market or a market for contingent claims on the raw material does not exist. (That the absence of a futures market for an input can provide an incentive for vertical integration is briefly noted in Crandall, 1968, p. 219.)

Granted the previous assumptions, then, downstream firms will have an incentive to purchase upstream firms in order to improve their ability to forecast the future price of the raw material input. Like Green (1974), Arrow (1975) finds an inherent tendency toward complete integration in the sense that no upstream firms will remain unintegrated. He further concludes, however, that no stable equilibrium can exist where there is vertical integration between the two stages of production but where ownership of upstream firms exists among more than one downstream firm. Thus, if vertical integration is permitted, the market structure of the intermediate-good and final-product industries must tend toward monopoly. This tendency is created by the assumption that the accuracy of raw material price prediction can always be improved by acquiring additional upstream firms. In effect, this assumption renders the downstream industry a natural monopoly, and the dynamic consequences follow tautologically. In such a world, a potential trade-off must exist between informational efficiency and competitive market structure. The net effect of such a trade-off on overall social welfare is a priori indeterminate (see Williamson, 1968).

The third study to appear since 1974 dealing with the role of uncertainty in firms' decisions to vertically integrate is Bernhardt (1977). This paper returns to the theme expressed in the earliest studies: the effect of vertical integration on the variability of demand for the intermediate product. Similar to these earlier studies, Bernhardt (1977) pointed out that fluctuations in demand impose costs on the firm by creating a need for inventory holdings, by necessitating the adoption of a flexible but relatively high-cost production technology, and so on. Here, however, the primary focus is on a derivation of the conditions under which integration (or contractual ties that bind individual buyers to a given seller) may be expected to reduce upstream demand variability. To begin with, three specific sources of variation in demand for the output of a given intermediate-good producer are identified, (1) random fluctuations in the aggregate demand for the product, (2) random fluctuations in the market shares of individual firms at the downstream stage, and (3) random fluctuations due to periodic shifting of sources of supply by firms at the downstream stage.

Bernhardt derived a model that compares the variances in quantity demanded over a fixed period of time that are experienced by input suppliers under two alternative regimes, zero and complete vertical integration. A comparison of these variances demonstrates that the effect of vertical integration on demand variability (and, hence, the existence of an incentive to integrate in this model) depends upon the specific source of demand variation in the absence of integration.

Three basic conclusions are established. First, vertical integration has no effect on demand variability that is due to random fluctuations in aggregate market demand. Thus, neither an incentive nor a disincentive to integrate is created by this source. Second, since vertical integration ties each downstream firm to a particular supplier, it will ordinarily increase the degree of variability experienced by firms at the upstream stage when demand fluctuations are due to random variations in the market shares of final-good producers. Here, then, there exists a disincentive for integration to occur. Finally, since vertical integration reduces or eliminates any incentive for downstream firms to shop, it is expected to eliminate demand variability resulting from intermediate-product users randomly shifting their sources of supply. Consequently, in this case, and in this case only, there is an incentive for vertical integration to occur in response to the intermediate-good producers' desire to reduce demand variability.

Next, Carlton (1979) constructed a model of vertical integration under uncertainty that is intended to reflect the business owner's often-expressed motive of assuring a more certain source of supply. To begin, Carlton described how a competitive market equilibrium is attained under conditions of uncertainty while characterizing the important features of such equilibria. In Carlton's model, production is not instantaneous. As a result, firms are forced to make their production decisions prior to observing their demand. In addition, prices are assumed to remain fixed during the market period and unit costs of production are assumed constant at c. Firms do not hold inventories, and any unsold units perish at the end of the market period. Thus, the firm faces two types of costs that it must trade off. First, if it runs out of the product and is unable to satisfy some customers, it bears the opportunity cost of lost sales. And second, if it produces more of the good than it can sell at the fixed price, it absorbs the production costs of the unsold units.

Buyers are assumed to know the price that the firm charges, but they do not know whether the firm has any unsold units left for sale at any particular time. Moreover, Carlton (1979) assumed that buyers do not engage in search. From the buyer's perspective, then, the product supplied by a particular firm has two relevant characteristics: price and probability that the good is available. Random demand for the individual firm's output is generated by a fixed number of buyers randomly selecting which firm they will choose to frequent.

In this model, the firm's expected profit depends upon the market price and the number of customers it can satisfy from its predetermined output. Consumers' utility depends upon price and the probability of finding the good available. Carlton (p. 193) drew iso-utility and iso-expected-profit curves in $1 - \lambda, p$ space (where $1 - \lambda$ is the probability that a buyer will be able to purchase the good and p is price). Both curves slope upward, but, for reasons described elsewhere, the iso-utility curves will be convex while the iso-expected-profit curves are concave. Assuming all customers and all firms to be identical, buyers will choose to frequent only those firms offering the highest level of satisfaction. And because of competition, all firms will be forced to bid up the utility levels that they offer until the expected profits are 0. Consequently, market equilibrium is attained at a tangency between the zero-profit curve and an iso-utility curve. This equilibrium is depicted in Figure 5.1 where $U(1 - \lambda, p) = \bar{U}$ is the highest iso-utility curve attainable along the 0 expected profit curve $E(\Pi) = 0$.

Carlton then described the important features of this equilibrium outcome. These are (1) the probability that a customer will find the product unavailable, λ, exceeds zero, (2) the market price exceeds the marginal cost of production (to cover the firm's average cost of unsold units), and (3) the total quantity supplied and quantity demanded are not necessarily equal. Given this characterization of competitive equilibrium under uncertainty, Carlton proceeded to analyze a more specific model in which this sort of behavior manifests itself across an intermediate-product market in order to examine the resulting incentive for vertical integration.

To do this, Carlton assumed two stages of production with random final good demand and a Leontief production technology at the downstream stage. The marginal cost of producing the intermediate-product is c. Any unused input must be discarded so that the producer of that input (an upstream firm or, with vertical integration, a downstream firm) loses c. There are L identical final product consumers, N_1 downstream firms, and N_2 upstream firms.

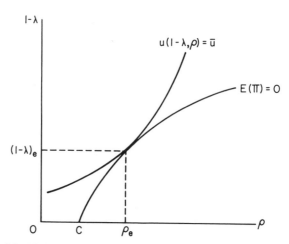

Figure 5.1 Market equilibrium price and probability of obtaining the good.

Each of the L consumers randomly frequents one downstream firm in each market period. The downstream firm attempts to obtain a unit of the intermediate good to satisfy the customer's demand. If this firm is vertically integrated, then it will first meet this demand with its own holdings of the input. Only when its internal supplies are depleted will it enter the intermediate-product market to satisfy its customer's demand. If the downstream firm is unable to obtain the input from the upstream firm that it randomly selects, then it is unable to satisfy its customer's demand for the final product.

In this model, there are two opposing forces influencing the decision of a downstream firm to produce a unit of the input (i.e., to vertically integrate). First, since the intermediate product is priced above its marginal cost ($p_e > c$), the firm saves $p_e - c$ per unit produced internally. At the same time, however, the firm stands to lose c on any units produced that go unsold. Given these two offsetting influences, the downstream firm will engage in *some* vertical integration (i.e., it will produce at least 1 unit of the input internally) if and only if

$$[1 - \Pr(0)]p_e > c \tag{5.1}$$

where $\Pr(0)$ is the probability that the firm will not be able to sell any output at all. The intuitive explanation of condition (5.1) is straightforward. By producing one unit of the intermediate product, the downstream firm's costs increase by c with certainty. At the same time, the firm's expected savings from not having to purchase that unit of the input in the intermediate-product market is the market price times the probability that

that first unit will be needed, $[1 - \Pr(0)]p_e$. If expected savings exceed costs, then some vertical integration will occur.

Given the definitions and assumptions of this model, the random process involved is binomial with probability $1/N_1$ and size L. Carlton showed that:

$$1 - \Pr(0) = 1 - \left(1 - \frac{1}{N_1}\right)^L \cong 1 - \exp\left(-\frac{L}{N_1}\right) \qquad (5.2)$$

Consequently, condition (5.1) becomes

$$\left[1 - \exp\left(-\frac{L}{N_1}\right)\right]p_e > c \qquad (5.3)$$

The left-hand side of this inequality varies directly with the customer-per-firm ratio at the downstream stage L/N_1. Or, for given L, this inequality is more likely to be satisfied the smaller the number of firms at the downstream stage. More importantly, since $p_e > c$ by Carlton's earlier analysis and since the bracketed term is close to unity even for relatively low customer-per-firm ratios, the above inequality will almost always hold. For $L/N_1 = 20$ (i.e., 20 customers per firm at the final-good stage), the bracketed term in expression (5.3) equals .999. For $L/N_1 = 5$, this term equals .993. This model indicates a strong incentive for some vertical integration to occur. The downstream firms employ their own input supplies to meet their "high probability" demand and purchase from upstream firms only to meet their "low probability" demand. Thus, with partial vertical integration, upstream firms fulfill an insurance role for the downstream producers.

Carlton further argued that the welfare effects of vertical integration in this model depend upon whether $N_1 > N_2$. If the number of firms at the downstream stage is larger than the number of firms at the upstream stage, then the upstream firms will be relatively large and, therefore, will be more efficient absorbers of risk. Since vertical integration involves some shifting of risk to the downstream producers, welfare will be reduced by vertical integration undertaken in this situation. On the other hand, if $N_1 < N_2$, then vertical integration improves welfare. Carlton's main point, however, is that private incentives to integrate can exist in either case. Consequently, we may observe vertical integration occurring even in situations in which markets are competitive and social welfare is reduced by such integration.

Up to this point, all of the studies that we have reviewed explicitly or implicitly assume risk neutrality on the part of the firms involved. The profit incentive to vertically integrate in these models stems from real resource savings that are realized by the firm as a result of either reducing risk or shifting it to other firms. In the presence of uncertainty, however, vertical integration may also be undertaken for diversification purposes if decision makers are risk averse. A recent paper by Perry (1982) and an earlier discussion by Warren-Boulton (1978, pp. 28–30) analyze the conditions under which vertical integration can be expected to result in a diversification of risk.

Assume that two firms exist at vertically related stages of production. Profits to the upstream firm are Π_u, and profits to the downstream firm are Π_D. These profits are assumed to be random variables due to the existence of some uncertainty that arises from a source to be specified later. The variances of these profits are σ_u^2 and σ_D^2, respectively. Then, if these two firms vertically integrate and no other effects of vertical integration are present, the variance of the profits of the integrated firm will be

$$\sigma_I^2 = \sigma_U^2 + \sigma_D^2 + 2\,\text{cov}(\Pi_U, \Pi_D) \tag{5.4}$$

where $\text{cov}(\Pi_U, \Pi_D)$ is the covariance between the profits of the two separate firms. Thus, the two firms will be able to diversify risk by vertically integrating if and only if $\text{cov}(\Pi_U, \Pi_D) < 0$.

The fundamental result obtained informally by Warren-Boulton (1978) and formally by Perry (1982) is that the sign of this covariance (and, hence, the existence of a diversification incentive to integrate) depends upon the specific source of uncertainty and, in some cases, the values of certain key parameters (the price elasticity of final-product demand, the elasticity of substitution at the downstream stage, etc.) At least four cases may be identified. Perry (1982) treated the first three, while Warren-Boulton (1978) briefly discussed all four of these cases.

First, if the uncertainty is due to random variations in the demand for the final product, then $\text{cov}(\Pi_U, \Pi_D) > 0$, and no incentive for vertical integration exists. Where $\text{cov}(\Pi_U, \Pi_D) > 0$, the overall variance of profits at the two stages is increased by vertical integration. Thus, there is a positive disincentive to integrate in this case. An increase in final-good demand increases Π_D and, at the same time, increases the derived demand for the intermediate product, causing Π_U to increase as well. A reduction in demand at the downstream stage has the opposite effect, causing the profits of both firms to fall. Thus, profits at the two stages will move in the same direction when the random element enters through variations in final-good demand. Integration does not lead to diversification in this case.

Second, where the intermediate product is employed in more than one final-good industry, random variations in the derived demands of these other industries will create an uncertain price for the input. In this case, some diversification probably will be achieved by vertical integration. An increase in the external demand for the intermediate product will increase the input's price and increase Π_U but (depending upon the relative magnitudes of the elasticity of substitution at the downstream stage and the elasticity of final good demand) will probably decrease Π_D. A reduction in external demand has the opposite effects. Consequently, it is likely that $\text{cov}(\Pi_U, \Pi_D) < 0$ in this case, and if so, the variance of profits can be reduced by vertical integration.

Third, random variations in the supply price of an input used in the production of the intermediate product are not likely to result in diversification opportunities. An increase in an input price at the upstream stage is likely to decrease both Π_U and Π_D, depending upon elasticities of substitution and demand at both stages. A decrease is likely to have the opposite effects. Thus, we expect $\text{cov}(\Pi_U, \Pi_D) > 0$ in this case, and vertical integration is not likely to reduce the variance of profits.

Fourth, random variations in the supply price of some other input employed at the downstream stage may create a diversification incentive for vertical integration. An increase in this other input's price will reduce Π_D. If this input is a substitute for the intermediate product in the production of final output, however, the substitution effect will outweigh the output effect, and the derived demand for the intermediate good will increase, causing Π_U to rise. Thus, if the two inputs are substitutes, $\text{cov}(\Pi_U, \Pi_D) < 0$, and the diversification incentive is present. If the two inputs are complements, however, $\text{cov}(\Pi_U, \Pi_D) > 0$, and no such incentive arises.

When one incorporates risk-averse behavior in the presence of uncertainty, incentives to vertically integrate may arise that are not related to the diversification motive. The following two sections examine such incentives.

Random Demand

Suppose that an intermediate product A is subject to monopoly supply.[1] This input is used in producing a final good B. We assume that the production of B is competitively organized. Moreover, we assume that the competitive producers face a random demand function. In particular, let the price of B be given by

$$\tilde{P}_B = P_B(\tilde{u}) \tag{5.5}$$

[1] This section and the next depend heavily upon Blair and Kaserman (1978b) and Blair and Kaserman (1982).

where the tilde denotes a random variable. The random variable u can enter this function additively with an expected value of zero, multiplicatively with an expectation of one, or in some more general way. Since u is random, the price to the competitive firm is a random variable with (subjective) probability density function $f(P_B)$ and mathematical expectation, $E[P_B] = \bar{P}_B$. For the industry, demand has the customary negative slope, $\partial P_B / \partial Q_B < 0$, but an increase in the random component will shift the demand, $\partial P_B / \partial u > 0$.

If the monopolistically produced intermediate good A subsequently is used in fixed proportions with other inputs to produce a unit of final output, we may assume, without loss of generality, that each unit of the final good B requires one unit of input A. Further, we assume that the *industry* marginal cost of converting a unit of A into a unit of B is constant, and is written as MC_C. Thus, for any given price of the intermediate good P_A, the industry marginal cost of producing B would be constant:

$$MC_B = P_A + MC_C. \tag{5.6}$$

This does not require that the marginal cost be constant for every firm; rather, the long-run supply function for the competitive industry is perfectly elastic at constant input prices. In fact, it is convenient to assume that MC_C is a U-shaped function because each downstream firm would be of indeterminate size if MC_C were constant for each firm.

In general, each competitive firm will attempt to maximize the expected utility of profit:

$$E[U(\Pi)] = E[U(P_B Q_B - (P_A + MC_C)Q_B)] \tag{5.7}$$

where E is the expectations operator and U is a von Neumann-Morgenstern utility function.[2] The competitive firm will produce its optimal output according to the first-order condition:

$$\partial E[U(\Pi)]/\partial Q_B = E[U'(\Pi)(P_B - P_A - MC_C)] = 0 \tag{5.8}$$

which can be written as

$$\bar{P}_B - P_A - MC_C = - \frac{cov[U'(\Pi), P_B]}{E[U'(\Pi)]} \tag{5.9}$$

where $cov[U'(\Pi), P_B]$ denotes the covariance between the marginal utility of profit and the final-good price P_B. For a derivation of this expres-

---

[2] An important, early contribution to the expected utility literature was provided by Milton Friedman and Leonard J. Savage (1948). A summary, with many important extensions, of the expected utility hypothesis applied to the firm was developed by Ira Horowitz (1970).

sion, see Baron (1970). Briefly, it follows directly from the definition of covariance:

$$E[XY] - E[X]E[Y] = \text{cov}[X, Y].$$

In this case, X is $U'(\Pi)$ and Y is P_B. For risk-neutral firms, the right-hand side of (5.9) is zero because $U'(\Pi)$ would be constant and the covariance between a constant and a random variable is zero. For risk averters (lovers), however, the right-hand side of (5.9) is positive (negative). Since the von Neumann-Morgenstern utility function increases monotonically, $U'(\Pi)$ is always positive as is its expectation. As P_B increases, say, the firm's profits will increase. For a risk-averse firm, however, the utility function is concave and $U'(\Pi)$ will decline. Thus, the covariance between price and marginal utility of profit is negative. Due to a convex utility function for a risk-loving firm, the covariance would be positive. Thus, the risk-averse firm's optimal output is less than that of the risk-neutral firm. As long as the costs and the probability distribution on price remain constant, the expected-utility maximizer will continue to produce the same output each period and to sell it at the market clearing price. We eschew the added complications that arise in connection with inventory problems by assuming that final good B is perishable—at least in the economic sense.

To see the impact of these results on the upstream monopolist, suppose that the industry demand for the final good takes the simple form

$$\tilde{Q}_B = a - bP_B(\tilde{u}) \tag{5.10}$$

which is random. The expected industry demand function is given by

$$E[Q_B] = \overline{Q}_B = a - bE[P_B] \tag{5.11}$$

Each of the competitive firms will produce so as to satisfy condition (5.9). Firms will enter or leave the industry if expected profit is too large or too small relative to the risk preferences of the marginal firm. By solving (5.9) for expected price and substituting into (5.11), we can obtain the derived demand for the monopolist's intermediate good:

$$Q_B = Q_A = a - b\left\{P_A + \text{MC}_C - \frac{\text{cov}[U'(\Pi), P_B]}{E[U'(\Pi)]}\right\} \tag{5.12}$$

We should note that the derived demand is not stochastic. Given all ex ante relationships, the demand for all inputs will be manifested *prior* to observing actual demand. Without any change in cost conditions or the distribution of demand for the final output, there will be no fluctuation in the demand for inputs. Thus, as can be seen in (5.12), there are no random variables in the derived demand function.

Whether an incentive exists for forward integration depends upon the relative attitudes toward risk of the competitive downstream producers and the monopolistic upstream supplier. We have assumed that all firms have homogeneous expectations or perceptions regarding the underlying probability distributions. The problem that we address here would surface in a somewhat different form if all firms had identical risk preferences but expectations were heterogeneous. The incentives and disincentives for vertical integration that we discuss further on, which are due to differences in risk preferences, would be similar for heterogeneous expectations. For completeness in a formal sense, we could examine the behavior of risk lovers, but such firms may simply be aberrational. An examination of first-order condition (5.9) reveals that risk lovers will have an expected profit below 0 in the long run. Thus, the risk lover may not survive very long. In any event, we shall confine our attention to risk aversion and risk neutrality. Our comparisons are cast in terms of risk averters, on the one hand, as opposed to risk-neutral firms on the other hand. This is purely for expositional convenience. All that is necessary for what follows is that there exists a difference in risk preferences.

First, suppose the monopolist and the competitive firms are *risk neutral*. There is neither an incentive nor disincentive for vertical integration under these conditions. The covariance terms in (5.9) and (5.12) disappear due to the linearity of the von Neumann-Morgenstern utility function. We are left with the stochastic analog of the deterministic case: the input monopolist is indifferent about integrating forward. This result can be demonstrated graphically. In Figure 5-2, the expected demand for final output is labelled $E[D]$ with the associated expected marginal revenue $E[MR]$. Given the marginal cost functions, MC_A and MC_C, the derived demand for input A is given by d with the associated marginal revenue mr. Note that $d = E[D] - MC_C$. The monopolist is indifferent about vertically integrating because his profit without integration equals the *expected* profit he can obtain through integration. For a risk-neutral firm, there is no risk premium. Thus, the firm is indifferent between a certain profit Π_0 and a random profit $\tilde{\Pi}$ such that its expected value equals Π_0. Without integration, the monopolist will price input A such that $Q_A = Q_{BM}$ is sold at a price equal to $P_{BM} - MC_C$, which yields a *deterministic* profit to the input monopolist of $(P_{BM} - MC_B)Q_{BM}$. If the monopolist were to vertically integrate, the optimal output would remain Q_{BM}, the *expected* price of the final output would be P_{BM}, and *expected* profit would be $(P_{BM} - MC_B)Q_{BM}$.

Alternatively, let us assume that the monopolist is *risk averse* and the competitive firms are *risk neutral*. In this event, the covariance terms in (5.9) and (5.12) again disappear. As a result, the monopolist selects the same price and output configuration as in the first case. What distin-

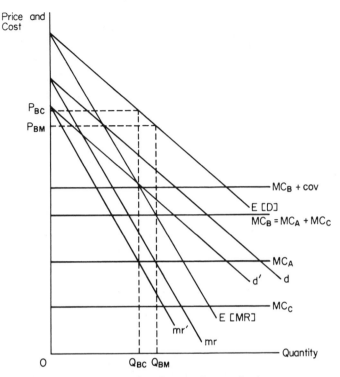

Figure 5.2 Effect on upstream profits of uncertainty at the downstream stage with all final-good producers equally risk averse.

guishes this case from the previous one is that when the monopolist and the competitive firms were all risk neutral, the monopolist was equally well off whether vertically integrated or not. In this instance, however, vertical integration would expose the upstream monopolist to risk, which is avoided by the risk averse firm. Assuming that the monopolist would still be risk averse after integration, the monopolist's response to this risk would be to lower expected profits through a reduction in output (price) if the firm behaves as a quantity (price)-setter. Thus, not only is there a lack of incentive for forward integration, but there is an actual disincentive.

Finally, suppose the monopolist is *risk neutral* while the competitive firms are *risk averse*. Thus, the covariance term in (5.9) is negative, which makes the right-hand side positive. The natural interpretation of this term is as an addition to marginal cost. In a sense, it represents the marginal (psychic) cost of bearing risk. Defining the Pratt-Arrow risk premium Z as being the amount that will satisfy

$$U(E[\Pi] - Z) = E[U(\Pi)],$$

one can show that the right-hand side of (5.9) equals dZ/dQ. This means that the covariance term divided by expected marginal utility of profit represents the incremental change in the risk premium as output changes. See Baron (1970, pp. 468–469). From the first-order condition (5.9), one may observe that this cost of bearing risk has the most important characteristic of all costs, that is, it is a payment that must be made to keep the firm's resources employed in that industry.

In this case, the monopolistic restriction of final-good output, which operates through the pricing of the intermediate good, is compounded by the output curtailment of the competitive final-good producers, which is due to their aversion to risk. Since the monopolist is not averse to risk, he or she would prefer a larger output of the final good. This can be seen by solving (8) for P_A as a function of Q_B and specifying the monopolist's profit function as:

$$\Pi_A = P_A(Q_A)Q_A - C_A(Q_A)$$

where $C_A(Q_A)$ is the total cost of producing A. Maximizing Π_A yields optimal Q_A as:

$$Q_A^* = \frac{a}{2} - \frac{b}{2}\,MC_C + MC_A - \frac{\text{cov}[U'(\Pi), P_B]}{E[U'(\Pi)]}$$

Vertical integration would eliminate the covariance term in this expression because of the monopolist's assumed risk neutrality. Thus, optimal production of the intermediate product would be

$$Q_A^{**} = \frac{a}{2} - \frac{b}{2}(MC_C + MC_A)$$

Clearly, since

$$\frac{\text{cov}[U'(\Pi), P_B]}{E[U'(\Pi)]} < 0, \quad Q_B^* < Q_B^{**}$$

One avenue by which an output expansion can be induced is through forward integration.

In Figure 5.2, the industry marginal cost in the case of risk neutrality was MC_B. With risk aversion, however, we must add the right-hand side of (5.9) to MC_B, which we have labelled $MC_B + \text{cov}$. For $MC_B + \text{cov}$ to be horizontal, one must assume an elastic supply of competitive firms with the same attitude toward risk. This assumption is relaxed in Figure 5.3, but does not lead to *qualitative* differences. With risk-averse downstream producers, the derived demand for the monopolist's input is d' rather than d, where $d' = E[D] - MC_C - \text{cov}$. Thus, the competitive

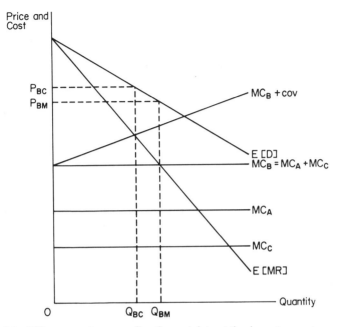

Figure 5.3   Effect on upstream profits of uncertainty at the downstream stage with varying degrees of risk aversion.

final-good producers would produce Q_{BC} in response to profit-maximizing price and output decisions of the input monopolist. Under our assumptions, Q_{BC} also represents the amount of the monopolized input A that is necessary. If the monopolist integrates forward, final output will be determined on the basis of MC_B rather than MC_B + cov because there is no cost of risk bearing to the risk-neutral monopolist. As a result, output will expand to Q_{BM} and the monopolist's expected profit will be $(P_{BM} - MC_B)Q_{BM}$ as opposed to the deterministic profit of $(P_{BC} - MC_B - \text{cov})Q_{BC}$.

In Figure 5.3, we relax the assumption that the supply of competitive risk bearers is perfectly elastic. In this case, as output expands, the marginal payment for bearing risk in the industry increases. Thus, the long-run marginal cost for the industry will be positively sloped and the inframarginal firms will enjoy some excess profit. The price and output without integration will be P_{BC} and Q_{BC}, whereas with vertical integration, price will fall to P_{BM} and optimal output will expand to Q_{BM}. Again, the result is due solely to having the firm most willing to bear risk actually do so. If both the monopolist and the competitive firms were risk averse, the same qualitative results would hold as long as there exists a difference in the degree of risk aversion with the upstream firm being relatively less risk averse.

Some Results For An Exponential Utility Function

As an example, we shall assume that the downstream competitive firms have von Neumann-Morgenstern utility functions of the exponential form

$$U_i(\Pi) = -e^{-\gamma_i \Pi} \tag{5.13}$$

where γ_i is the Pratt-Arrow index of absolute risk aversion.[3] Of course, the expectation of $\tilde{P}_B$ is $\bar{P}_B$ for all firms and the variance of $\tilde{P}_B$ is given by σ^2. Assuming that the probability distribution of price is normal, maximizing expected utility of profit is equivalent to maximizing the certainty equivalent:

$$Z = \bar{P}_B Q_B - (P_A + MC_C)Q_B - \tfrac{1}{2}\gamma_i Q_B^2 \sigma^2 \tag{5.14}$$

where $Q_B^2 \sigma^2$ is the variance of profits. For each firm, the first-order conditions for an interior maximum require

$$\frac{\partial Z}{\partial Q_B} = \bar{P}_B - (P_A + MC_C) - \gamma_i \sigma^2 Q_B = 0 \tag{5.15}$$

This, of course, can be solved for Q_B, which is equivalent to the derived demand of the ith firm since one unit of A is needed for each unit of B:

$$Q_B = \bar{P}_B - (P_A + MC_C)/\gamma_i \sigma^2 \tag{5.16}$$

The total derived demand confronting the input monopolist can be written as

$$Q_B = [\bar{P}_B - (P_A + MC_C)]R \tag{5.17}$$

where $R = \sum_{i=1}^{n} 1/\gamma_i \sigma^2$.

Due to the quantity-setting behavior of the competitive firms, the monopolist is in a deterministic setting and wants to maximize profit. Since Q_A equals Q_B on our assumption, the monopolist must set its price, P_A, such that

$$\Pi = Q_B P_A - C(Q_A) \tag{5.18}$$

---

[3] If $\gamma_i = \gamma_j$ for all i, j, we have the analog of the case depicted in Figure 5.2. In contrast, if $\gamma_i \neq \gamma_j$, the case depicted in Figure 5.3 is appropriate for comparison.

is maximized. Substituting (5.17) into (5.18) and differentiating with respect to P_A yields the first-order condition

$$\frac{d\Pi}{dP_A} = (\bar{P}_B - P_A - MC_C)R - P_A R + MC_A R = 0 \qquad (5.19)$$

which can be solved for the optimal price P_A^*:

$$P_A^* = \tfrac{1}{2}(\bar{P}_B - MC_C + MC_A) \qquad (5.20)$$

Substituting (5.17) and (5.20) into (5.18) yields a simple expression for maximum profit:

$$\Pi^* = R/4(\bar{P}_B - MC_C - MC_A)^2 \qquad (5.21)$$

The impact upon the monopolist of an increase in absolute risk aversion by the downstream firms can be obtained by examining the sign of

$$\frac{\partial \Pi^*}{\partial \gamma_i} = \frac{\partial \Pi^*}{\partial R}\frac{\partial R}{\partial \gamma_i} = \frac{1}{4}(\bar{P}_B - MC_C - MC_A)^2(-1/\gamma_i^2\sigma^2) \qquad (5.22)$$

which is, of course, negative.[4] Thus, an increased aversion to risk of the downstream firms or an increased price variance have a deleterious impact upon the profits of the input monopolist.[5] Since these effects are due to actions of the competitive firms, the input monopolist can remove these influences by forward integration.

A Contractual Alternative

With random final product demand and risk aversion at the downstream stage, the risk-neutral (or relatively less risk-averse) input monopolist can achieve results that are similar to vertical integration through the use of consignment sales. That is, the upstream firm may contract to buy back

---

[4] The qualitative result obtained in equation (5.22) is unchanged by assuming that all downstream firms have the same attitude toward risk. In that event, R would equal $n/\gamma\sigma^2$ and the impact of a change in risk attitude is given by

$$\frac{\partial \Pi^*}{\partial \gamma} = \frac{\partial \Pi^*}{\partial R}\frac{\partial R}{\partial \gamma} = \tfrac{1}{4}(\bar{P}_B - MC_C - MC_A)^2(-n/\gamma^2\sigma^2)$$

which is, of course, negative.

[5] The impact of a change in the variance is obtained by signing

$$\frac{\partial \Pi^*}{\partial \sigma^2} = \frac{\partial \Pi^*}{\partial R}\frac{\partial R}{\partial \sigma^2} = \tfrac{1}{4}(\bar{P}_B - MC_C - MC_A)^2\left(-\sum_{i=1}^{n} 1/\gamma_i(\sigma^2)^2\right)$$

which is negative. A reduction in the variance is broadly analogous to increased information in the model analyzed by Arrow (1975).

any unused units of the monopolized input. Such an agreement has the effect of shifting the risk of demand variations to the upstream firm which is more readily equipped to absorb it.

Consignment sales, however, may not be entirely equivalent to vertical integration. The downstream producers will have an incentive to overpurchase the monopolized input in order to have enough units available to satisfy their low probability, high level of demand. In other words, they will purchase more of the input than would a risk-neutral firm. Consequently, the input monopolist will be in a position of almost always having to buy back some of the input. These return purchases will, of course, increase costs at the upstream stage. Therefore, we may expect to observe consignment sales only where managerial diseconomies discourage ownership integration.

Implications

The analysis of this section has several implications for public policy and the theory of the firm. First, we have seen that if the monopolist is risk neutral (averse) and the competitive downstream industry is characterized by risk aversion (neutrality), then there is a positive incentive (disincentive) for forward integration.

A second implication of this analysis is that divergent risk preferences between firms at the various stages of production provide an additional incentive for (or against) vertical integration. (Even if risk preferences are the same, similar results will be obtained if the firms' perceptions of the (subjective) probability densities are different.) Recognizing the existence of such an incentive contributes to a more complete understanding of the firm's internalization decisions. Moreover, the benefits to the monopolist of vertical integration in the presence of uncertainty may lead to vertical integration when managerial diseconomies would preclude integration in the absence of uncertainty.

Finally, the welfare effects of vertical integration or consignment sales spawned by the incentive to accept or avoid risk in this model are clearly positive. Vertical integration (or the lack thereof) that allows risk to be borne by the firm or firms with the least aversion to it results in increased output of the final good and a reduced expected price. In these circumstances, it can be viewed as a market-perfecting mechanism. By reallocating risk to those firms most willing to accept it, vertical integration acts as a hybrid form of insurance without the moral hazards. Given positive welfare effects under these circumstances, public policy should be formulated so that integration emanating from this incentive is not discouraged.

Random Input Prices

The influence of random input prices[6] can also be analyzed in the context of the fixed-proportions model. Specifically, we shall suppose that inputs x_1 and x_2 are combined in fixed proportions for the purpose of producing a final good Q. With no loss of generality, we may assume that each unit of final good Q requires one unit of x_1 and one unit of x_2. The now-classic example concerns the production of knives, which requires one handle and one blade per knife. See Marshall (1930, Book V, Chapter 6). Thus, the production function for the final good is assumed to be

$$Q = \min(x_1, x_2)$$

The supply of x_1 and the production of Q are presumed to be competitively organized. In contrast, we assume that the production and subsequent sale of x_2 is monopolized. Moreover, we shall permit the production of x_2 to be characterized by linear homogeneity. In combination with constant input prices, linear homogeneity will result in constant marginal (and average) costs of producing x_2. In this situation, the input monopolist's problem is to select the price and output that will maximize its profits.

With the downstream industry in a position of long-run competitive equilibrium, profits at the downstream stage are exhausted and the following relationship will hold

$$PQ - p_1 x_1 - p_2 x_2 = 0 \tag{5.23}$$

where P is final output price and p_1, p_2 are input prices. Then, the inverse derived demand function for x_2 may be obtained by solving (5.23) for p_2,

$$p_2 = P \frac{Q}{x_2} - p_1 \frac{x_1}{x_2}$$

or

$$p_2 = P - p_1$$

since $Q/x_2 = x_1/x_2 = 1$ along the Pareto ray. Since the x_1 market is competitive by assumption, we have

$$p_2 = P - c_1 \tag{5.24}$$

where c_1 is the constant marginal cost of x_1. If the inverse demand curve

---

[6] The influence of random input prices has been considered in a fairly general context by Blair (1974). Random input price along with delayed production is also employed by Arrow (1975).

for final output is linear and given by $P = a - bQ$, then (5.24) may be rewritten as

$$p_2 = a - bx_2 - c_1 \qquad (5.25)$$

since $Q = x_2$.

The profit function of the x_2 monopolist is

$$\Pi_{x_2} = p_2 x_2 - c_2 x_2 \qquad (5.26)$$

where c_2 is the constant marginal cost of producing x_2. Substituting (5.25) into (5.26), we obtain

$$\Pi_{x_2} = (a - c_1 - c_2)x_2 - bx_2^2 \qquad (5.27)$$

which is the profit function of the x_2 monopolist that reflects the derived demand for x_2. (Note that $a - c_1 - c_2 > 0$ must hold if production is to occur at all.) Maximization of (5.27) yields the optimal quantity for the upstream monopolist to produce

$$x_2^* = (a - c_1 - c_2)/2b \qquad (5.28)$$

Substituting (5.28) into (5.25) and (5.27), we obtain the optimal price for the monopolized input

$$p_2^* = \tfrac{1}{2}(a - c_1 + x_2) \qquad (5.29)$$

and maximized profits are

$$\Pi_{x_2}^* = (a - c_1 - c_2)^2/4b \qquad (5.30)$$

This process is depicted in Figure 5.4. The final-good demand function is given by D. If we denote the competitive price of x_1 by p_1, which equals the constant marginal cost of x_1, then the derived demand as expressed in equation (5.24) is denoted by d and the associated marginal revenue by mr. The constant marginal cost of x_2 is indicated by MC_2. Profit maximization requires that the input monopolist produce x_2^* ($= Q^*$) units of input 2. These will be sold to the final-good industry at p_2^* and the input monopolist's profits will be $(p_2^* - MC_2)x_2^*$. Since we have assumed for expositional convenience that one unit of x_2 is required for each unit of Q, Q^* units of the final good will be produced and will be sold to consumers at the industry marginal cost, which is the sum of p_2^* and p_1.

The combination of random input prices and risk aversion causes a reduction in total output relative to the deterministic output level. We shall assume that the final-good producers must commit themselves to specific production levels before they are able to observe the input prices that each will have to pay. Generally, it is unreasonable to expect production to be instantaneous. Consequently, it would not seem uncommon for

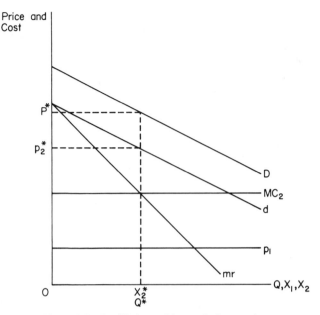

Figure 5.4 Equilibrium with certain input prices.

a firm to plan future output levels on the basis of orders for future delivery and some probabilistic notions of what the input prices will be. Of course, the firm will be aware of the actual costs only when the necessary inputs are purchased.

The profit function for each of the final-good producers is now assumed to be given by

$$\Pi_Q = PQ - \tilde{p}_1 x_1 - p_2 x_2 \qquad (5.31)$$

where the tilde on p_1 denotes a random variable. We assume that $\tilde{p}_1 = p_1(u_1)$ where u_1 is a random variable with probability density $f(u_1)$ and expectation

$$E[u_1] = \int_\alpha^\infty u_1 f(u_1)\,du_1 = 0$$

The lower bound, α, is put on u_1 to ensure that p_1 will not be negative. By definition, then, profit is a random variable, and profit maximization is no longer relevant. We shall assume that the firm maximizes the expected utility of profit.

If the competitive final-good producers are risk neutral, then each is concerned only with expected profit. The variance of profit is of no concern and each firm seeks to maximize expected profit. Consequently, the stochastic analog of the results presented above can be obtained by substituting the expected value of p_1 for p_1 in (5.28), (5.29), and (5.30). The derived demand for x_2 is unaffected and the monopolistic supplier of x_2 consequently is unaffected by the introduction of uncertainty.

In contrast, if the final-good producers are risk averse, then the input monopolist may be quite concerned about random input prices. Specifically, a risk-averse firm does not like uncertainty and will attempt to avoid it completely. Failing that, the firm will attempt to mitigate the impact of uncertainty. As a result, its behavior will be cautious. In the context of random input prices, caution leads the risk-averse firm to curtail its employment below that level where the value of the marginal product equals the expected input price. This can be seen in the first-order conditions for the final-good producer as derived in Blair (1974), in which the objective function is $E[u(\Pi)] = E[u(PQ - p_1x_1 - p_2x_2)]$. In effect, the firm's perceived cost of employing an additional unit of the input exceeds the expected price of that unit to account for the risk premium that risk bearing requires. For our purposes, the qualitative effect of this behavior is to make the expected price of x_1 irrelevant by itself. In its place, we must consider $\bar{p}_1 + r$, which includes the adjustment for risk. The effect of this adjustment, then, is to reduce the x_2 monopolist's optimal output and profits since

$$dx_2^*/dp_1 = -1/2b < 0 \tag{5.32}$$

and

$$d\Pi_{x_2^*}/dp_1 = -(a - c_1 - c_2)/2b < 0 \tag{5.33}$$

Since $x_2^* = Q^*$, final output is also reduced, as a result of uncertainty.

The process involved is shown in Figure 5.5. This figure was constructed so that D and MC_2 are precisely the same as in Figure 5.4. In addition, $\bar{p}_1$ is identical to p_1. The impact of risk aversion on the part of the downstream producers is to reduce the derived demand for x_2. This follows because the inverse derived demand is now given by

$$p_2 = P - p_1 - r$$

due to risk aversion.

It is clear that the derived demand is not itself subject to randomness although it is obviously affected by uncertainty. The reason that the derived demand is nonstochastic is that the expected utility maximizing firms at the downstream stage will not alter their output decisions unless the probability distribution of the random input price changes. As a conse-

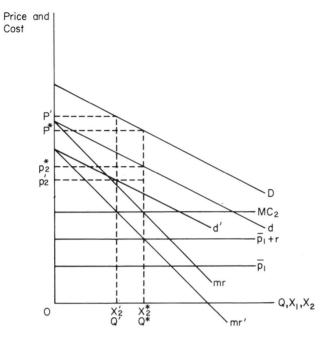

Figure 5.5   Effect of random input prices on upstream profits.

quence, the derived demand is d' rather than d and the relevant marginal revenue is mr'. The input monopolist will maximize its profits by supplying x_2' at a price of p_2'. Since x_2' is less than x_2^*, the quantity of final good falls to Q'. This leads to a corresponding price increase to P' from P^*. The profits of the input monopolist fall to $(p_2' - MC_2)x_2'$ relative to the profits associated with risk neutrality. There can be little doubt that the input monopolist would be better off if the deleterious effect of uncertainty in the price of the other input could be removed.

One way of coping with uncertainty is to vertically integrate forward into the final-good stage. If the input monopolist elects to produce the final good, he can behave in a risk-neutral fashion. Thus, the appropriate cost of input 1 will be $\bar{p}_1$. As a result, output will expand to Q^* and the input monopolist's expected profits will be higher.

A Contractual Alternative

Tie-in sales provide another means of dealing with the uncertainty caused by random input prices. If the input monopolist were to employ a

tying arrangement, we should expect him to purchase x_1, at the competitive, albeit random, price $\bar{p}_1$. The final-good producers will then buy one
unit each of x_1 and x_2 for each unit of output they intend to produce. The
input monopolist may undertake to supply x_1 at $\bar{p}_1$ in every period. In this
case, the price of the tied x_1, x_2 package would be $p_2^* + \bar{p}_1$. Thus, the
final-good producers will be insulated from uncertainty and the derived
demand for x_2 will be restored to d in Figure 5.5. The input monopolist's
profits on the sale of the x_1, x_2 package will now be random,

$$\Pi_{x_1,x_2} = (p_2^* - MC_2)x_2^* + (\bar{p}_1 - \tilde{p}_1)x_1$$

with expectation

$$E[\Pi_{x_2,x_2}] = (p_2^* - MC_2)x_2^*$$

which exceeds the certain profits available without tying.

Implications

The welfare effects of vertical integration or tying in this instance are
decidedly positive: the output of the final good expands and its price falls.
Consequently, the consumers are better off. There is no reason for the
final-good producers to resist the input monopolist's offer since it reduces
uncertainty. The suppliers of x_1 should also be better off because the
demand for x_1 expands. Thus, there are no losers, only gainers.

Input monopolists are not exposed to uncertainty when they only supply x_2. By electing to supply x_1 at a fixed price, they expose themselves to
the risks associated with random prices. If the input monopolists are risk
neutral, that is, indifferent to risk, they should behave as described in the
preceding paragraph because they will be better off, on the average. If
they are less risk averse than the competitive final-good industry, then
their qualitative behavior will be similar. Vertical integration, or tying,
still makes economic sense and has positive welfare effects. Only if the
input monopolist is more risk averse than the competitive final-good industry will vertical integration or the tying arrangement be eschewed.[7]
Thus, if vertical integration or tying arises in response to this sort of
uncertainty, social welfare is enhanced.

---

[7] Being the sole producer of x_2, the monopolist may well experience a lower variance of
profits than do the individual firms at the competitive downstream stage simply because of
the larger size of this firm. Thus, an incentive to tie might arise from the superior risk-
spreading capabilities of the upstream monopolist. Advantages of an informational nature
may also lead to equivalent results (see Arrow, 1975). All of these cases utilize tying to
realize efficiency gains, that is, the price of the bundled good is less than the sum of the
prices of the component goods purchased separately. Telser (1979) examined alternative
incentives for tying in cases where such efficiencies do not arise.

The need for tying in this case is clear. The final-good producer will be only too happy to buy the requirements for x_1 from the x_2 producer whenever the actual price of x_1 exceeds the mean. But when the actual price is below $\bar{p}_1$, the attempt will be made to purchase x_1 on the open market. The x_2 producer wants to use the gains from sales at prices above $\bar{p}_1$ to offset the losses when prices are below $\bar{p}_1$, and thus, will have to insist on purchase commitments for all prices of x_1.

6

Vertical Integration without Contractual Alternatives

Several potential incentives for vertical control exist where there is no obvious contractual alternative to ownership integration. In this chapter we discuss four such incentives: single stage rate-of-return regulation, monopsony power over input purchases, price discrimination in intermediate-good sales, and persistent long-run disequilibrium. In situations in which these incentives are operative (and where opposing incentives do not dominate), we should expect to observe common ownership of the relevant stages of production. Given the original market situation, we shall see that the welfare effects of adopting an integrated structure may be positive, negative, or indeterminate according to which particular incentive is providing the motivation for vertical integration.

Regulation and Vertical Integration

Assume that a single firm at the final stage of production experiences declining average costs throughout a range of production sufficiently great to satisfy the entire market demand. Then, efficiency of resource utilization requires that a monopoly position be obtained by the firm. Uninhibited by regulatory constraint, such a firm could be expected to exercise full monopoly power in the output market and would not have any incentive to integrate backward into input supply unless some market imperfection existed at the intermediate-product stage. But if the firm is subjected to effective rate-of-return regulation, a unique incentive for vertical integration into the capital equipment supply industry is created.

David Dayan (1973) has shown that if regulation is restricted to the final stage of production and the firm is permitted to integrate upstream into the equipment supply industry, then it is possible for the firm to circumvent the effect of the regulatory constraint completely by transfer pricing the internally supplied intermediate product above its marginal production cost. As he notes, the possibility of avoiding the regulatory constraint by adopting a vertically integrated structure has been discussed by previous authors. (See Adelman, 1949; Edwards, 1953; Sheahan, 1956; Irwin and McKee, 1968; and Kahn, 1971.) By integrating backward, the regulated firm can increase its profits and, at the same time, avoid input distortions of the Averch-Johnson variety. These results can be easily shown with the following model.

Let

K = capital input;

L = labor input;

$Q = Q(K, L)$ = final output production function, which is assumed to be continuous and twice differentiable with $Q(K, 0) = Q(0, L) = 0$ and $\partial Q/\partial K, \partial Q/\partial L > 0$;

c = constant unit and marginal cost of the capital input;

r = annual rate of interest;

w = wage rate;

$P = P(Q)$ = inverse demand for final output, with $dP/dQ < 0$;

$R(K, L) = P[Q(K, L)]Q(K, L)$ = total revenue function; and

s = the allowed rate of return permitted by the regulatory authority, where $s > r$.

Without integration, the monopolist's profit function will be given by

$$\Pi_N = R(K, L) - wL - rcK \tag{6.1}$$

since we are assuming the market for the capital input to be perfectly competitive. If the monopolist integrates into the production of the capital input and transfer prices this input to itself at p, its profit function will be

$$\begin{aligned} \Pi_I &= R(K, L) - wL - rpK + rpK - rcK \\ &= R(K, L) - wL - rcK \\ &= \Pi_N \end{aligned} \tag{6.2}$$

That is, with perfectly competitive input markets and no regulation, no incentive for backward integration exists. The application of effective rate-of-return regulation to the final stage of production, however, changes this result.

Rate-of-return regulation limits the amount that the firm can earn on its invested capital. The ratio of gross revenues less operating expenses to the value of the firm's capital input (the rate base) is limited to a given percentage amount; that is, for the nonintegrated firm the constraint is given by

$$\frac{R(K, L) - wL}{cK} \leq s \qquad (6.3)$$

Here, the rate base cK is given by the competitive price of the capital equipment times the quantity of capital employed by the firm. As has been shown elsewhere, maximization of (6.1) subject to (6.3) results in the well-known Averch-Johnson effect: the firm employs more than the cost-minimizing quantity of capital for any given level of output. (See Averch and Johnson, 1962; Takayama, 1969; Bailey and Malone, 1970; Zajac, 1970; and Baumol and Klevorick, 1970.)

If the regulated firm integrates into the production of K and the regulatory authority permits the capital equipment to be valued at the transfer price p in the rate base, the regulatory constraint becomes

$$\frac{R(K, L) - wL}{pK} \leq s \qquad (6.4)$$

Dayan's (1973) results then follow from proof of the following two propositions.

Proposition 6.1   Rate-of-return regulation with valuation of the capital input at the transfer price provides an incentive for vertical integration into equipment supply.

Proof   Effective regulation implies that constraints (6.3) and (6.4) will be met as equalities. In the absence of vertical integration, (6.3) implies

$$R(K, L) - wL = scK \qquad (6.5)$$

Substituting (6.5) into (6.1), the nonintegrated regulated firm's profit function is

$$\Pi_N = (s - r)cK \qquad (6.6)$$

With vertical integration, (6.4) implies

$$R(K, L) - wL = spK \qquad (6.7)$$

Substituting (6.7) into (6.2), the vertically integrated regulated firm's profit function is

$$\Pi_I = (sp - rc)K \qquad (6.8)$$

Adding and subtracting scK to (6.8), the integrated firm's profits are

$$\Pi_I = (s - r)cK + (p - c)sK > \Pi_N \qquad (6.9)$$

for $p > c$.

Proposition 6.2   The vertically integrated regulated firm will employ inputs in efficient proportions; that is, it will minimize the cost of producing a given level of output.

Proof   Cost minimization requires that capital and labor be employed in proportions that result in the ratio of their marginal physical products being equal to the ratio of their prices. With vertical integration and rate-of-return regulation, the firm attempts to maximize (6.2) subject to (6.4). The integrated firm's control variables for this problem are the quantities of both inputs and the transfer price of the capital equipment. The Lagrangian for this problem is

$$\mathcal{L} = R(K, L) - wL - rcK + \lambda[spK - R(K, L) + wL], \quad (6.10)$$

with first-order conditions

$$\frac{\partial \mathcal{L}}{\partial K} = \frac{\partial R}{\partial K} - rc + \lambda sp - \lambda \frac{\partial R}{\partial K} = 0 \quad (6.11)$$

$$\frac{\partial \mathcal{L}}{\partial L} = \frac{\partial R}{\partial L} - w - \lambda \frac{\partial R}{\partial L} + \lambda w = 0 \quad (6.12)$$

$$\frac{\partial \mathcal{L}}{\partial p} = \lambda sK = 0 \quad (6.13)$$

Since $s, K > 0$, equation (6.13) implies $\lambda = 0$, that is, the rate-of-return constraint is rendered ineffective. Then, (6.11) and (6.12) yield

$$\frac{\partial Q/\partial K}{\partial Q/\partial L} = \frac{rc}{w} \quad (6.14)$$

and the proof is complete.

The above propositions demonstrate that single stage rate-of-return regulation provides a profit incentive for the firm to integrate backward into equipment supply and that the effect of such integration is to render the regulation of the final stage ineffective. As Dayan (1973) has shown, effective regulation in the presence of vertical integration requires that each stage of production be regulated individually or, equivalently, that both the transfer price of the intermediate product and the final output price be subjected to rate-of-return controls. Assuming that the original regulation of the final stage of production is desirable and that additional regulation of the capital equipment stage is not costless, it is apparent that, in the absence of any efficiencies of integration, social welfare is damaged by internalization that derives from this particular incentive. In the presence of efficiencies of integration, social welfare will be improved by the adoption of a vertically integrated structure only if the cost savings attributable to vertical integration exceed the costs of extending the regulatory network to the intermediate product stage.

Monopsony and Vertical Integration

The theory of vertical integration by a monopsonistic buyer of an input has not received a great deal of attention in the literature. Wu (1964, pp. 118–119) showed that, if a monopsonist completely integrates backward (i.e., if the firm supplies its entire requirements of the input that was formerly purchased), then employment of the input will expand. This expansion leads, in turn, to an expanded output and reduced price for the final product. Thus, if such integration does occur, social welfare will be unambiguously improved by it. Wu did not, however, examine the profit incentive that the monopsonist has to acquire its input suppliers, since that was not the purpose of his paper (which is summarized in Chapter 5). Consequently, no theory is provided to predict whether or under what circumstances such vertical integration will, in fact, occur.

Martin Perry (1978c) provided a rigorous theory of vertical integration under monopsony. Perry's analysis, however, allows for partial vertical integration by the monopsonist and is, as a result, fairly complex. The basic profit incentive of a monopsonist to integrate backward can be demonstrated much more simply by comparing maximized profits under zero integration (where the monopsonist purchases its entire requirements of the input on the market) with maximized profits under complete integration (where the monopsonist purchases none of the input but produces its entire requirements internally). Using this approach, we can demonstrate Perry's (1978c) fundamental results.

We show that, ignoring any costs of acquiring upstream input suppliers, the monopsonist's maximized profits are always greater under complete vertical integration. We then go on to show that the combined profits of the monopsonist and its upstream suppliers may be either larger or smaller with complete vertical integration. Thus, whether or not complete vertical integration will actually occur under monopsony depends upon the profitability of input suppliers without integration and the ability of the monopsonist to acquire these suppliers at less than the total upstream profits through strategic purchasing behavior.

To analyze the incentive of a monopsonist to integrate backward, we make use of the following notation and assumptions:

P = final output price which is assumed to be constant to the monopsonist;

χ = quantity of the monopsonized input used by the monopsonist;

$Q = Q(\chi)$ = final output production function which is assumed to be continuous and twice differentiable with $dQ/d\chi > 0$ and $d^2Q/d\chi^2 < 0$;

$p = p(\chi)$ = inverse supply function of input χ to the monopsonist, and

$C(\chi)$ = total industry costs of producing input χ where we assume that $dC/d\chi > 0$ and $d^2C/d\chi^2 > 0$.

We assume that the producer of Q faces a perfectly competitive output market but is a monopsonist in the purchase of input χ. The supply curve of input χ is assumed to slope upward because the production of this input requires some limited resource, that is, χ is characterized by "Ricardian" increasing costs. Therefore, even though the χ market is assumed to be competitive and thus prices its output at industry marginal cost, there will be no long-run entry to drive price downward to industry average cost. Rents can continue to exist at the upstream stage.

Using the above notation and assumptions, the nonintegrated monopsonist faces a profit function that is given by

$$\Pi_N = \int_0^\chi P\,\frac{dQ}{d\chi}\,d\chi - p(\chi)\chi \qquad (6.15)$$

In the absence of vertical integration, the input supply industry, which we assume is competitive, will have a profit function that is given by

$$\Pi_s = p\chi - C(\chi) \qquad (6.16)$$

If the monopsonist completely integrates backward into the production of χ, its profit function will be

$$\Pi_I = \int_0^\chi P\,\frac{dQ}{d\chi}\,d\chi - C(\chi) \qquad (6.17)$$

and the input supply market will cease to exist. We can now investigate the incentive to integrate by comparing maximized profits under the two alternative market regimes.

Under the no-integration regime, profit maximization on the part of individual input suppliers, each of which views p as a constant, results in a supply price for input χ that is equal to the industry's marginal cost of producing χ. Thus, maximization of (6.16) results in

$$p = dC/d\chi \qquad (6.18)$$

Now, substituting (6.18) into (6.15), we differentiate Π_N with respect to

χ and set the result equal to 0 to obtain the first-order condition for profit maximization by the monopsonist

$$P \frac{dQ}{d\chi} = \frac{dC}{d\chi} + \chi \frac{d^2C}{d\chi^2} \tag{6.19}$$

Equation (6.19) implicitly defines χ_N^*, the optimal quantity of input χ for the monopsonist to employ. Substituting this value of χ back into equation (6.15), we obtain the monopsonist's maximized profit with no integration

$$\Pi_N^* = \int_0^{\chi_N^*} P \frac{dQ}{d\chi} d\chi - \frac{dC}{d\chi}\bigg|_{\chi_N^*} \chi_N^* \tag{6.20}$$

where $dC/d\chi|_{\chi_N^*}$ is the upstream industry's marginal cost of producing χ evaluated at χ_N^*.

Turning now to the vertically integrated firm, maximization of (6.17) yields the first-order condition

$$P \, dQ/d\chi = dC/d\chi \tag{6.21}$$

Equation (6.21) implicitly defines χ_I^*, the optimal employment level for input χ under complete vertical integration. Comparing (6.19) and (6.21), it is clear that $\chi_I^* > \chi_N^*$ since $dQ/d\chi$ is decreasing in χ in Knightian Region II. [Both Wu (1964) and Perry (1978c) show that $\chi_I^* > \chi_N^*$.] Substituting χ_I^* into (6.17), we obtain the integrated firm's maximized profit

$$\Pi_I^* = \int_0^{\chi_I^*} P \frac{dQ}{d\chi} d\chi - C(\chi_I^*) \tag{6.22}$$

We are now in a position to determine the profitability of vertical integration under monopsony. Ignoring for the moment any costs involved in the acquisition of productive capacity at the upstream stage, it will be profitable for the monopsonist to adopt a vertically integrated structure if and only if

$$\Pi_I^* - \Pi_N^* > 0 \tag{6.23}$$

Substituting from (6.20) and (6.22) and simplifying,

$$\Pi_I^* - \Pi_N^* = \int_{\chi_N^*}^{\chi_I^*} P \frac{dQ}{d\chi} d\chi - C(\chi_I^*) + \frac{dC}{d\chi}\bigg|_{\chi_N^*} \chi_N^* > 0 \tag{6.24}$$

so condition (6.23) is satisfied. With zero costs of acquiring upstream firms, complete backward integration by a monopsonist always improves the firm's profits. This result is shown graphically in Figure 6.1.

In this figure, the upstream industry's average total cost curve for pro-

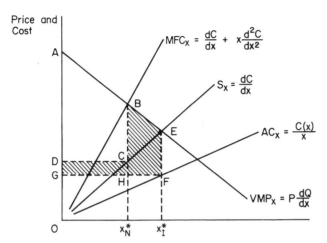

Figure 6.1 Profitability of backward integration by a monopsonist with zero acquisition costs.

ducing the input χ is given as $AC_\chi = C(\chi)/\chi$. Above AC_χ, we have graphed the industry's marginal cost curve which, under competition, corresponds to the industry supply curve, $S_\chi = dC/d\chi$. This supply curve is the average cost of purchasing χ to the nonintegrated, nonprice-discriminating monopsonist. Above S_χ we have graphed the marginal factor cost curve MFC_χ, which indicates the marginal cost to the nonintegrated monopsonist of increasing the employment of χ. Since the total cost of employing χ is $p(\chi)\chi$, $MFC_\chi = p + \chi(dp/d\chi)$. And since $p = dC/d\chi$, we have $MFC_\chi = dC/d\chi + \chi(d^2C/d\chi^2)$, as is shown in the graph. Finally, we have graphed the downward sloping value of the marginal product curve, $VMP_\chi = P(dQ/d\chi)$.

In the absence of vertical integration, the monopsonist will employ χ_N^* units of the input χ as determined by the intersection of VMP_χ and MFC_χ at point B. This intersection corresponds to the equality indicated in the first-order condition (6.19). At this level of employment, the monopsonist's total revenue is given by $\int_0^{\chi_N^*} P(dQ/d\chi)d\chi$, which is equal to the area $AB\chi_N^*0$. Total cost is $p\chi$, which is equal to the area $DC\chi_N^*0$. Consequently, the nonintegrated monopsonist's maximized profits are given by the area $ABCD$ in the graph.

With complete vertical integration, the optimal employment of χ increases to χ_I^* as determined by the intersection of VMP_χ and $dC/d\chi$ at point E. This intersection corresponds to the first-order condition (6.21). Here, total revenue is area $AE\chi_I^*0$, and total cost is area $GF\chi_I^*0$. Thus, the fully integrated firm's maximized profits are given by the area $AEFG$ in the graph. An equivalent result would be obtained if the monopsonist were able to practice first-degree price discrimination in purchasing input x.

The increase in profits that results from vertical integration by the monopsonist is given by the shaded area $DCBEFG$ in the graph. This area corresponds to the expression on the RHS of the equality in (6.24). Areas corresponding to each of the three terms within this expression may be seen in the graph. The term $\int_{\chi_N^*}^{\chi_1^*} P(dQ/d\chi)d\chi$ corresponds to the area $\chi_N^* BE\chi_1^*$. The term $dC/d\chi|_{\chi_N^*} \chi_N^*$, which is added to the first term, corresponds to the area $ODC\chi_N^*$. And the term $C(\chi_1^*)$, which is subtracted from these, corresponds to the area $OGF\chi_1^*$. That $\Pi_1^* - \Pi_N^* > 0$ is clearly shown in the graph since $\chi_N^* BE\chi_1^* + ODC\chi_N^* > OGF\chi_1^*$.

Figure 6.1 is also useful in demonstrating the two separate sources of increased profits discussed by Perry (1978c, pp. 568–569). What Perry terms the "rent effect" is shown as area $DCHG$ and is that portion of the upstream industry's preintegration rents that is captured by the monopsonist through backward integration. The second source of increased profits arises from the elimination of the efficiency loss of underemployment of the monopsonized input in the absence of vertical integration. We label this source the "efficiency effect," and it is shown as area $HBEF$ in the graph. Clearly, the relative sizes of these two effects depends upon the slope of the VMP_χ curve. As VMP_χ becomes more elastic (or as $d^2Q/d\chi^2$ approaches 0) the efficiency effect becomes larger and the rent effect becomes smaller. The sum of the profits resulting from the rent effect and the efficiency effect provides the total increase in profits from vertical integration under monopsony as shown by the shaded area in the graph.

This increase in profits measures the gross benefits to the monopsonist of acquiring productive capacity at the upstream stage. To determine whether such acquisition will actually occur, however, we need to compare these gross benefits with the costs of acquisition. That is, we need to measure the net benefits of vertical integration by the monopsonist, where net benefits are defined as gross benefits minus acquisition costs. Only if net benefits are positive will complete vertical integration occur.

Under competition, the cost of acquiring all existing firms at the upstream stage should equal the combined maximized profits of those firms under the no-integration market regime. Substituting (6.18) into (6.16), maximized profits of the input supply industry are

$$\Pi_S^* = \frac{dC}{d\chi}\bigg|_{\chi_N^*}\chi_N^* - C(\chi_N^*)$$

$$= \left[\frac{dC}{d\chi}\bigg|_{\chi_N^*} - \frac{C(\chi_N^*)}{\chi_N^*}\right]\chi_N^* \tag{6.25}$$

Using (6.20), (6.22), and (6.25), the net benefits of complete backward integration are given by

$$\Pi_I^* - (\Pi_N^* + \Pi_S^*) = \int_{\chi_N^*}^{\chi_I^*} P \frac{dQ}{d\chi} d\chi - [C(\chi_I^*) - C(\chi_N^*)] \qquad (6.26)$$

which may be either positive or negative since $C(\chi_I^*) > C(\chi_N^*)$. Figure 6.2 helps to clarify this result.

In this figure, we have reproduced AC_χ, S_χ, MFC_χ, and VMP_χ from Figure 6.1. The net benefits from vertical integration (6.26) are shown in Figure 6.2 as the difference between the shaded area *BEFH* and the double-shaded area *GHIJ*. These net benefits may be seen to be the gross benefits represented by the shaded area in Figure 6.1 minus the preintegration profits of the input supply industry that are given by the area *DCIJ* corresponding to Π_S^* in (6.25).

Further insight may be gained by interpreting the difference between the two shaded areas of Figure 6.2 in terms of the individual components of equation (6.26). In the graph, $\int_{\chi_N^*}^{\chi_I^*} P(dQ/d\chi)d\chi$ is given by the area $\chi_N^* BE\chi_I^*$, $C(\chi_I^*)$ is area $OGF\chi_I^*$, and $C(\chi_N^*)$ is $OJI\chi_N^*$. From (6.26), then, net benefits are $\chi_N^* BE\chi_I^* + OJI\chi_N^* - OGF\chi_I^* = BEFH - GHIJ$.

Since the difference between these two areas cannot, in general, be signed on a priori grounds, the incentive for a monopsonist to engage in complete vertical integration will depend upon the actual shapes of the two relevant functions, $C(\chi)$ and VMP_χ. That is, the incentive to internalize the supply of the monopsonized input completely will depend upon the industry cost function of that input and the production and demand func-

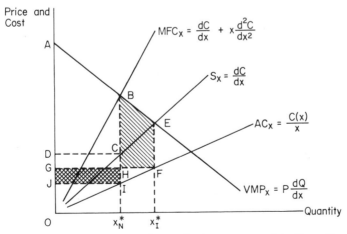

Figure 6.2 Profitability of backward integration by a monopsonist with positive acquisition costs.

tions for final output. Perry (1978c) demonstrated, however, that, even where complete backward integration is not profitable, some degree of vertical integration will always improve the nonintegrated monopsonist's profits. In that case, the optimal degree of vertical integration (defined as the fraction of upstream firms acquired) will depend upon the costs of acquiring additional input suppliers.

Price Discrimination and Vertical Integration

A firm possessing monopoly power in an input market may often sell its output to downstream firms with divergent elasticities of derived demand for its product, particularly in situations in which these downstream firms sell their output in separate geographic or product markets. Variations in final-product demand elasticities across downstream markets can easily lead to differences in elasticities of derived demand for the monopolized input. Such variations may be attributable to geographic disparities between the input and output markets or to the employment of a given input by more than one downstream industry.

Under such circumstances, it is well known that the upstream monopolist can increase profits by practicing price discrimination, selling the intermediate product at a lower price to downstream firms with relatively high demand elasticities. To be successful, however, such discrimination must be accompanied by some mechanism to prevent arbitrage of the intermediate product between downstream firms. Vertical integration by acquisition of relatively high-elasticity downstream firms (where elasticity is calculated at the optimal monopoly price to each firm) constitutes one such mechanism.[1] Through merger with these firms, the upstream monopolist can effectively eliminate incentives for arbitrage at the downstream stage. The increased profits available from effectuating the price-discrimination scheme may be shared by the parties to the merger so that both the input monopolist and the high-elasticity downstream firms can realize net gains from integration. In effect, vertical integration is used to achieve a consonance of objectives between the parties to the merger regarding the price to be paid for the intermediate product by the lower-elasticity (nonintegrated) firms.

Suppose we have an intermediate-product monopolist that sells its out-

---

[1] This incentive for vertical integration was first mentioned by Wallace (1937). It is also discussed in Stigler (1951), Edwards (1953), Burstein (1960), Williamson (1971), and Warren-Boulton (1977). Recent extensions of this incentive are provided in Gould (1977) and Perry (1978a).

put A to two competitive final-good industries that use input A to produce final products x and y. (This treatment is adapted from Gould, 1977.) For simplicity, we assume that 1 unit of A is required for each unit of x and each unit of y produced. To abstract from other potential incentives for vertical integration, we also assume that all other inputs are competitively supplied, that random elements are absent, that transaction costs are 0, and that all markets are always in long-run equilibrium. Then, if the derived demands for input A on the part of the x and y industries exhibit different price elasticities, the intermediate-good monopolist will want to practice third-degree price discrimination between these markets charging a higher price for A in the relatively inelastic market, say, y. Whether such discrimination is feasible, however, depends upon the ability of the input monopolist to prevent arbitrage.

If arbitrage is costlessly preventable, Figure 6.3 depicts the equilibrium outcome. To keep this figure as simple as possible, we assume that the marginal cost of producing input A, MC_A, and the marginal costs of transforming a unit of A into a unit of x and a unit of y, MC_{Tx} and MC_{Ty}, respectively, are all constant. Then, given final-product demands of D_x and D_y, the derived demands for input A in the two markets will be $D_x - MC_{Tx}$ and $D_y - MC_{Ty}$, with marginal revenues of $MR_x - MC_{Tx}$ and $MR_y - MC_{Ty}$. Equating marginal cost to marginal revenue in each market (no horizontal summation of marginal revenues is necessary with constant marginal cost), the price-discriminating, intermediate-good monopolist will sell A_x to industry x and A_y to industry y, charging P_{Ax} and P_{Ay} per unit in these two markets, respectively. Thus, with successful price discrimination, we will have the following set of equalities:

$$MC_A = MR_x - MC_{Tx} \quad \text{and} \quad MC_A = MR_y - MC_{Ty} \quad (6.27)$$

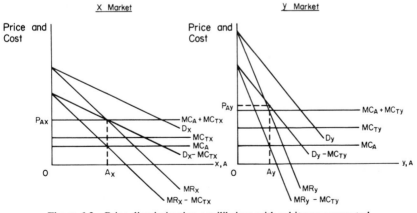

Figure 6.3 Price-discriminating equilibrium with arbitrage prevented.

Now, suppose that arbitrage cannot be prevented so that the price differential between the x and y markets cannot be sustained on the open market. By vertically integrating into the x industry, the input monopolist can completely monopolize that final-good industry, thereby removing the sales of A to x from the market. Then, the integrated monopolist will face a marginal revenue of MR_x and marginal costs of $MC_A + MC_{Tx}$ in the x market. The y market will remain as before. Consequently, the monopolist that integrates into the relatively high-elasticity market only will maximize profits by allocating input A to these two markets so that the following set of equalities holds:

$$MC_A + MC_{Tx} = MR_x \quad \text{and} \quad MC_A = MR_y - MC_{Ty} \quad (6.28)$$

which is equivalent to (6.27). Thus, vertical integration into the relatively high elasticity market yields results that are identical to those achieved under price discrimination.

Finally, vertical integration into both the x and the y markets will lead to the following set of equalities:

$$MC_A + MC_{Tx} = MR_x \quad \text{and} \quad MC_A + MC_{Ty} = MR_y \quad (6.29)$$

which is equivalent to both (6.27) and (6.28). Consequently, the input monopolist can achieve results that are equivalent to successful price discrimination by vertically integrating into either the high-elasticity market or both markets. Profits will be the same under all three strategies.

When the downstream firms or industries may be segmented into n groups with differing elasticities of derived demand (where $n > 2$), Perry (1978a) showed that, in the absence of any costs of integration, the upstream monopolist can increase its profits by complete vertical integration into the $n - 1$ firms or industries that exhibit elasticities of derived demand that are higher than the elasticity of derived demand of the nth firm or industry. Thus, given the ability to identify differing demand elasticities for its product, the input monopolist calculates the optimal price to charge each customer or group of customers as though it could prevent arbitrage and successfully practice price discrimination. Actually unable to prevent arbitrage, however, the monopolist orders its customers from 1 to n by optimal price with $p_1^* < p_2^* < \ldots < p_n^*$ (where p_i^* is the optimal price for the ith customer group if price discrimination could be carried out). The input monopolist then integrates completely into the $n - 1$ customer groups with relatively low optimal prices (or, equivalently, relatively high demand elasticities).

Once the input monopolist is completely integrated into the $n - 1$ customer groups, its profits will be identical to those that could be earned through price discrimination with arbitrage prevented. Moreover, further vertical integration into the nth customer group will leave the monopolist's profits unaffected. Consequently, the input monopolist's profits will be the same if: (a) it can practice price discrimination and prevent arbitrage, (b) it integrates completely the $n - 1$ customer groups with relatively high-demand elasticities, or (c) it integrates completely all n customer groups.

Perry (1978a) also shows that the process of integrating successively higher-elasticity customer groups may be optimally terminated at less than $n - 1$ groups if the upstream firm is a dominant firm (as opposed to a monopolist) in the input supply industry. Here, the supply curve of the nondiscriminating competitive fringe provides a constraint on the dominant firm's ability to increase price to the low-elasticity (nonintegrated) customer groups. In this model, the optimal number of customer groups to integrate will vary inversely with the level of the competitive input supply schedule.

Obviously, an analogous result would be obtained with pure monopoly at the upstream stage if one introduced increasing costs of extending the firm's control to additional customer groups into the model. Managerial diseconomies of integrating downstream industries (which, we would imagine, is not an unlikely occurrence) would provide a brake on the process of acquiring additional customer groups, stopping the process short of the $n - 1$th group. In this case, the input monopolist would acquire additional customer groups as long as the increase in profits resulting from the enhanced ability to discriminate in price outweighed the increase in the cost of internalizing an additional industry. Moreover, to the extent that the incremental managerial diseconomies of the individual customer groups do not follow the same ordering as the optimal prices to these groups, (also a not-unlikely occurrence), the input monopolist may find it optimal to integrate a lower-elasticity group before integrating a higher-elasticity group. Also, it is clear that, with managerial diseconomies, social welfare might be improved relative to the integrated case if arbitrage could be costlessly prevented and price discrimination carried out without vertical integration.

In cases where arbitrage at the downstream stage can be prevented, however, the input monopolist might still opt for the vertical integration approach because of the threat of antitrust prosecution under the Robinson-Patman Act. Price discrimination is a per se violation of the antitrust statutes, but market exchange at divergent prices is necessary to establish its existence. By removing the lower-priced transactions from the market, vertical integration effectively eliminates the threat of detection. As a result, existing prohibitions may encourage a greater amount of vertical integration than would be forthcoming in their absence and may encourage integration in situations in which efficiency considerations would dictate simple price discrimination.

In addition to the above incentive to integrate forward, Crandall (1968) has shown that a desire to implement a price-discrimination scheme may also provide an incentive for backward integration. When the intermediate product (repair parts) is also purchased by subsequent consumers of the final good (automobiles) in inverse relation to their elasticity of demand for the final product, consumption of the intermediate product may serve as a metering device that yields information concerning relevant demand elasticities. Given the existence of monopoly power at the final-good stage, price discrimination may be facilitated by recognition of the complementarity that obtains between the final and intermediate products. Also, profits may be increased through vertical integration by lowering the price of the final product for which demand is relatively elastic, and raising the price of the complementary intermediate good for which demand is relatively inelastic. The profit incentive to implement such a scheme is larger the higher the proportion of sales of the intermediate product relative to sales of the final product.

Essentially, integration in these circumstances involves acquisition of productive capacity for a product for which consumption is highly complementary to the firm's original monopolized product and for which the price elasticity of demand is below that of the original product. Such acquisition increases the firm's profits by allowing the exercise of a given degree of monopoly power in a relatively inelastic market. Obviously, the feasibility of this strategy critically depends on the condition of entry into the less-elastic market. Whether such a practice constitutes price discrimination in the traditional sense is not clear, but the potential incentive for vertical integration is apparent.

Finally, since the exercise of price discrimination either directly or through vertical integration results in an expansion of output in the relatively elastic markets and a contraction of output in the relatively inelastic markets, its net effect on social welfare (holding constant the monopoly power on which it is based) is indeterminate on a priori grounds. Consequently, vertical integration that stems from a desire to implement price discrimination may or may not be desirable from a public policy point of view.

Vertical Integration and Downstream Disequilibrium

Under certain assumptions, it has been shown that an input monopolist has an incentive to vertically integrate forward and completely monopolize a formerly competitive downstream industry (Chapter 4). In practice, however, we can observe partial integration: the input monopolist transfers some of its output to subsidiary fabricators and the rest is sold to independent firms at the downstream stage. Casual empiricism suggests that such partial integration is more prevalent than total integration. Although there are a number of reasons why an input monopolist might choose to limit the degree of forward integration, in this section we shall focus upon one that has escaped mention in the prior literature: integration to achieve long-run competitive equilibrium in the downstream industry.

The fact that disequilibrium at the downstream stage creates an incentive for vertical integration is of interest for three reasons: (1) this incentive is consistent with and may help to explain the phenomenon of partial integration, (2) most existing analyses of vertical integration implicitly assume long-run equilibrium of the downstream industry, and (3) important public policy implications flow from the analysis we shall present.

Our analysis is motivated by the widely recognized failure of markets to adjust instantaneously to a position of long-run competitive equilibrium. Due to a general lack of perfect information and an overall inertia of productive resources, industries that are essentially competitively structured and whose firms view market price as an exogenous parameter may often remain in a disequilibrium state for protracted periods. In such industries, vertical integration may act as a surrogate for entry to achieve or accelerate a movement to long-run competitive equilibrium at the downstream stage.

We show that the incentive for such entry via integration transcends a simple desire on the part of the upstream firm to capture the transitional economic profits that materialize during the period of adjustment. Rather, this incentive derives from a profit motive to have equilibrium restored in the downstream industry. Moreover, since the theory demonstrates that the intermediate-product monopolist has more to gain from entry than does any other firm (either within the industry or outside the industry), we should not be surprised to see the former enter (integrate) more rapidly than the latter. Thus, while vertical integration has often been viewed as a mechanism for increasing barriers to entry (see Chapter 3), it is seen here to be a likely method for overcoming them as well.

We shall examine a monopolist that sells its output to a competitively structured distribution industry. For each unit sold at retail, the distributor must buy a unit from the manufacturer. Thus, the production of retail units is according to fixed proportions. Distribution services are produced competitively, transactions costs are assumed to be absent, and the final-product demand curve is assumed to be nonstochastic. In these circumstances, disequilibrium provides the only potential incentive for vertical integration.[2]

Prices and Outputs

The initial equilibrium for the manufacturer is depicted in Figure 6.4 where D_1 represents the retail demand for the product, MC_D represents the marginal cost of distributing the product, and $d_1 = D_1 - MC_D$ is the derived demand for the manufacturer's product. Given the marginal cost of producing the product MC_P, the manufacturer will produce X_1^1 units of output (where mr_1 equals MC_p) and charge a price of p_1^1. The retail price will be equal to p_1^1 plus the marginal cost of distribution (MC_D), which is labelled P_1.

A typical retailer is depicted in Figure 6.5, where MC and AC represent the firm's marginal and average costs, respectively. These cost curves incorporate the manufacturer's profit-maximizing price p_1^1. Competition among the retailers leads to a retail price of P_1, which is equal to p_1^1 plus MC_D. Each distributor will sell Q_1 units at retail.

Now, suppose that the final-good demand shifts to D_2 from D_1. If the number of distribution firms in the industry could change instantaneously, the manufacturer's derived demand would shift to $d_2 = D_2 - MC_D$. The new price and output (p_1^2 and X_1^2) are determined by equating mr_2 with MC_P. Retail customers then pay P_2 for Q_2 ($=x_1^2$) units of output. But the number of firms is not apt to change instantaneously due to a variety of potential frictions.

From the manufacturer's perspective, the problem is to determine the optimal wholesale price during the period of disequilibrium. In the absence of entry, the new derived demand curve will be d_3 in Figure 6.4, which is obtained by subtracting mc_D from D_2. Now, mc_D represents the industry marginal cost of distribution services when the number of distributors is held constant. Notice that, as we have drawn mc_D, we have

---

[2] The assumption of fixed proportions at the downstream stage eliminates the incentive to integrate provided by producer substitution away from the monopolized input (Burstein, 1960; Vernon and Graham, 1971). Use of the monopolized input by a single competitive downstream industry renders vertical integration for purposes of price discrimination unlikely (Perry, 1978a). Competition in distribution eliminates the incentive to integrate that arises from successive monopoly (Spengler, 1950). The absence of transaction costs obviates any incentive to avoid such costs by internalization of the transaction (Coase, 1937; Williamson, 1971). And the lack of uncertainty concerning final-product demand eliminates any incentives to integrate arising from information or risk considerations (Arrow, 1975; Blair and Kaserman, 1978; Carlton, 1979).

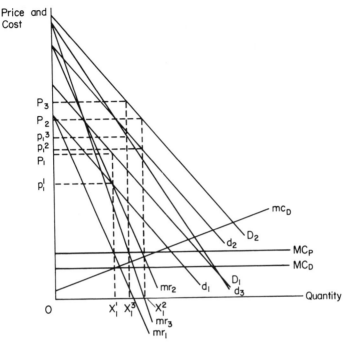

Figure 6.4 Profitability of forward integration with downstream disequilibrium.

implicitly assumed that exit barriers as well as entry barriers exist. That is, due to the presence of industry-specific capital, firms will produce along their marginal cost curves below average total cost. In this case, $mc_D < MC_D$ for $X < X_1^1$, and the upstream firm could thus profit from disequilibrium at the downstream stage during demand slumps. If, on the other hand, we assumed an asymmetry to entry and exit barriers with free exit but blocked entry, mc_D would equal MC_D up to X_1^1 and would exceed MC_D above X_1^1. Such an assumption, then, would result in a kink in the derived demand schedule d_3 at X_1^1. As each distributor attempts to expand its retail output along the MC curve in Figure 6.5, diseconomies are encountered. Collectively, the efforts of the distributors lead to a short-run marginal cost of distribution function like mc_D as opposed to the long-run marginal cost of distribution MC_D.

Given the disequilibrium, then, the manufacturer's derived demand becomes d_3, which is equal to $D_2 - mc_D$. In order to maximize profit, the manufacturer will select an output of X_1^3 where mr_3 equals MC_P. The wholesale price will be p_1^3 and the corresponding retail price will be P_3. Of course, P_3 is equal to p_1^3 plus mc_D evaluated at an output of Q_3.

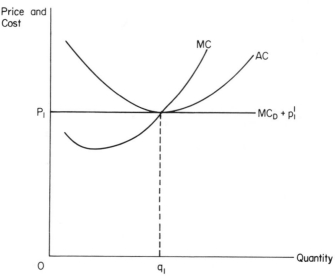

Figure 6.5   Equilibrium for the individual retailer.

Profit Comparisons

For the manufacturer, the long-run profit inherent in the final-good demand D_2 is equal to $(p_1^2 - MC_P)X_1^2$. In the absence of opportunities for price discrimination, this is the maximum profit that any industry organization could generate. In long-run equilibrium, the competitive downstream distributors will earn 0 excess profits. Their price to retail customers P_2 will be equal to the wholesale price p_1^2 plus the per-unit cost of the distribution services.

During the disequilibrium period, however, there are too few distributors. As a consequence, the manufacturer's profit is reduced to $(p_1^3 - MC_P)X_1^3$. In Figure 6.4, it is apparent that this profit is smaller than that associated with long-run equilibrium. But this is a general result. The long-run profit must exceed the short-run profit under these circumstances because short-run output is restricted, due to an inefficiency. This inefficiency, created by downstream industry disequilibrium, bears a broad similarity to the successive monopoly problem, but it is analytically distinct.

During the disequilibrium period, the marginal cost exceeds the average cost of distribution as each distributor expands output beyond q_1 in Figure 6.5. This short-run excess profit provides the signal that additional resources should be invested in this distribution industry.

Incentives for Entry

Due to the presence of quasi-rents, all resource owners outside this distribution industry share the traditional incentive for entry. The excess profits enjoyed by the firms in that industry act as a lure to attract resources from outside the industry. This includes the manufacturer of the monopolized product. If the choice is made to vertically integrate forward, the manufacturer will share in the transitional profits that persist until the distribution industry's capacity is expanded to a total output level of Q_2. There is nothing unique about this incentive. But the manufacturer has an additional incentive for vertically integrating forward. Disequilibrium in distribution has an adverse effect upon the derived demand of the monopolist, which reduces profits on the manufacture and sale of the product. Consequently, there is an *additional* incentive to vertically integrate forward into distribution to restore long-run equilibrium. (See this chapter's appendix for formal proof of this result.)

Our basic point is that this incentive to enter the distribution industry is analytically distinct from the normal incentive to enter an industry that is earning excess profits. This incentive is not shared with other potential entrants. Thus, in the absence of entry at the downstream stage (or, in a dynamic model, if entry proceeds at too slow a pace), the upstream firm will integrate forward to restore the final-product industry to a position of long-run equilibrium. Consequently, since the upstream monopolist's profits are increased more than outside firms' profits by the process of entry, any entry barriers that might exist at the downstream stage are more likely to be surmounted by this particular firm than by any other firm.

Moreover, the incentive to integrate that derives from this source will persist only until long-run equilibrium is restored at the downstream stage. Consequently, vertical integration that is due to disequilibrium is likely to stop far short of a complete monopolization of the final-product market. Partial integration, as opposed to complete integration, can be expected to result. Partial integration by an upstream monopolist results in what has been termed "dual distribution." The practice has come under antitrust attack (see "Justice Takes Aim at Dual Distribution," *Business Week*, July 7, 1980, pp. 24–25). As we will show, such attack may have perverse social welfare effects.

The Welfare Effects

Although some economists and many public policy analysts are suspicious of vertical integration, in this instance their concerns should be allayed by the generally positive welfare consequences; vertical integration has occurred to overcome imperfect information, inertia, or some other entry-retarding friction. The final result is a larger output (Q_2 as opposed to Q_3) and a lower price (P_2 rather than P_3) for the final consumer. This provides a rather unambiguous and positive welfare effect.

Monopolists, of course, are not motivated by undiluted altruism. We have seen that they benefit from rapid entry at the distribution stage. Since total industry output also expands with entry, however, the retail consumers are seen to benefit as well. In this case, the interests of the manufacturer and the consumer coincide. If entry into the distribution industry is not forthcoming or if it is too slow, the upstream monopolist can increase profits and simultaneously improve overall social welfare by forward vertical integration.

We should expect that the positive welfare effects stemming from forward integration in this example will be greater: the larger the shift in final good demand, the greater the price elasticity of this demand, the steeper the slope of downstream firms' marginal cost curves, and the slower the rate of entry by other firms.

Policy Implications

We have demonstrated a basic incentive for an input monopolist to integrate forward into a competitive customer industry when a disequilibrium situation creates positive economic profits at the downstream stage. It is important to realize that, unlike the normal channels of entry, forward integration in this case is not simply an attempt to capture those profits that exist during the adjustment back to a state of long-run competitive equilibrium. Rather, integration provides a strategy for permanently removing the distorting effect that disequilibrium has on the derived demand for the intermediate product.

Without attempting to describe the particular circumstances that might have created the situation of disequilibrium in the final-product market or the barriers that might exist to retard entry into this market by outside firms, we simply note that an upstream monopolist could well be in a particularly strategic position to detect the existence of short-run profits at the downstream stage and to eliminate such profits through forward integration. Therefore, a policy that discourages forward integration in this situation may serve only to prolong disequilibrium and thereby maintain an unnecessarily high output price.

Finally, the preceding theory of vertical integration, in conjunction with the alternative theories that exist, provides some (albeit somewhat loose) justification for the common notion that public policy should treat vertical integration via merger as being fundamentally different from vertical integration via internal expansion. Mueller (1969) appeared to advocate the use of such a distinction. For a critical discussion of this notion, see McGee and Bassett (1976). If the primary motivation for observed vertical integration is either transaction-cost savings or entry (both clearly beneficial to social welfare), then output at the final stage of production will need to expand. While particular cases can be imagined in which such expansion might occur following, or in conjunction with, integration via merger (McGee and Bassett, 1976), one would generally expect to observe de novo entry into the downstream stage in these cases. Vertical integration that arises from the variable proportions incentive, on the other hand, may lead to either an expansion or contraction of output at the final stage, with the latter appearing to be the more likely (Hay, 1973). In addition, since the input monopolist is capable of devaluing the assets of downstream producers by exercising a price-cost squeeze on them (Schmalensee 1973, p. 449), it is likely that the merger or acquisition route will be the least-cost path for the monopolist to take in acquiring productive assets at the downstream stage.

Since the social welfare effects of vertical integration of this latter sort are a priori indeterminate (Warren-Boulton, 1974), the above distinction does not justify a hostile attitude toward all vertical integration that occurs through merger or acquisition. It does, however, justify the exercise of somewhat more caution in the treatment of this particular category of vertical integration.

Appendix: A Formal Analysis of Vertical Integration and Downstream Disequilibrium

In order to isolate the incentive for vertical integration with which we are concerned, we assume a situation in which no other incentive to integrate can exist. Thus, we assume that an intermediate-product monopolist sells its output x_1 to a single competitively structured downstream industry that employs this product in fixed proportions with one other input x_2 in the production of final output, Q. The x_2 market is assumed to be competitive, transaction costs in transferring inputs to the downstream industry are assumed to be absent, and the final-product demand curve is assumed to be known with certainty.

To admit the possibility of long-run disequilibrium at the downstream stage, we assume that the production function for Q, while exhibiting fixed input proportions, does not generate constant costs. Thus, we assume that the production function at the final stage is given by

$$Q = \min\left[\frac{x_1}{\alpha_1(Q)}, \frac{x_2}{\alpha_2(Q)}\right] \qquad (A.1)$$

where $\alpha_1(Q)$ and $\alpha_2(Q)$, the production coefficients, satisfy the condition that

$$\frac{\alpha_1(Q_0)}{\alpha_2(Q_0)} = \frac{\alpha_1(Q_1)}{\alpha_2(Q_1)} = C, \qquad (A.2)$$

for all Q_0, $Q_1 > 0$, where C is a positive constant and $Q_0 \neq Q_1$. The production function given in equation (A.1) then exhibits the property that input ratios remain constant regardless of the input price ratio or the output level selected and is characterized by increasing, constant, or decreasing returns to scale as $d\alpha_i(Q)/dQ \lesseqgtr 0$, $i = 1, 2$. Assuming $\alpha_i(Q)$ is U-shaped, then the firm's average cost curve will also be U-shaped.

Assuming that production occurs along the Pareto ray (i.e., assuming that inputs are not wasted), the quantity of each input employed at any given level of output will be given by

$$x_i(Q) = \alpha_i(Q)Q$$

$i = 1, 2$; and the quantity of each input demanded will respond to changes in the quantity of final output produced by

$$\frac{dx_i}{dQ} = \alpha_i + Q\frac{d\alpha_i(Q)}{dQ} \qquad (A.3)$$

along this ray. As shown by Ferguson (1969), it is necessary that $dx_i/dQ \geq 0$ in order to preclude the economically meaningless possibility that a greater output could be produced with a lesser quantity of both inputs. Finally, condition (A.2) implies that

$$\frac{d\alpha_1(Q)/dQ}{d\alpha_2(Q)/dQ} = \frac{\alpha_1(Q)}{\alpha_2(Q)} \qquad (A.4)$$

Now, with the x_2 and Q markets competitive and the x_1 market monopolized, the upstream firm that controls x_1 has at least two alternative strategies available. First, the firm may simply sell x_1 to downstream producers at the profit-maximizing price, in which case the input monopolist's profits will be

$$\Pi_u = p_1(x_1)x_1 - c(x_1) \qquad (A.5)$$

where p_1 denotes the price of the intermediate product x_1 and $c(x_1)$ denotes the total cost of supplying this product. Or second, the firm may integrate forward and monopolize the final-product industry, in which case the integrated monopolists's profits will be

$$\Pi_I = P(Q)(Q) - c(x_1) - p_2 x_2 \qquad \text{(A.6)}$$

where P is the price of the final product and p_2 is the (competitive) price of the input x_2.

Given these two alternatives, the x_1 monopolist will have no incentive to integrate forward into the production of Q as long as the downstream industry is in a position of long-run competitive equilibrium. If, however, an increase in the demand for Q pushes final output price above the minimum point on the downstream firms' long-run average cost curve and if corrective entry is not immediate, then the x_1 monopolist will, in fact, have an incentive to integrate forward. Moreover, unlike the normal incentive to enter an industry where price exceeds average cost, this incentive is separate from and in addition to any motive provided by the lure of short-run economic profits that might be captured at the downstream stage. That is, the upstream firm's profits will remain higher even *after* the downstream industry is returned to a position of long-run competitive equilibrium. These statements can be shown to be true by proof of the following two propositions.

Proposition A.1 If the downstream industry is in a position of long-run competitive equilibrium, then $\Pi_u = \Pi_I$.

Proof Long-run competitive equilibrium implies that final-output price is equated to average cost at the downstream stage, or

$$P = (p_1 x_1 + p_2 x_2)/Q$$

This, in turn, implies that the derived demand price for the monopolized input is

$$p_1 = (PQ - p_2 x_2)/x_1 \qquad \text{(A.7)}$$

From (A.5) and (A.6), $\Pi_u = \Pi_I$ requires that

$$p_1(x_1)x_1 - c(x_1) = PQ - c(x_1) - p_2 x_2 \qquad \text{(A.8)}$$

Substitution of (A.7) on the LHS of (A.8) completes the proof.

Proposition A.2 If price exceeds average cost at the downstream stage, $d\alpha_1(Q)/dQ > 0$ must hold (i.e., decreasing returns to scale), and $\Pi_u < \Pi_I$.

Proof   Substituting from (A.5) and (A.6), $\Pi_u < \Pi_I$ requires that

$$p_1(x_1)x_1 - c(x_1) < P(Q)Q - c(x_1) - p_2x_2 \qquad (A.9)$$

In the absence of vertical integration, the downstream industry will continue to exist, and industry profits will be given as

$$\Pi_D = PQ - p_1x_1 - p_2x_2$$

which, in disequilibrium, will not equal zero. Assuming that the disequilibrium is created by an exogenous increase in final product demand, $\Pi_D > 0$. Maximization of profits by the individual firms in this industry, each of which views P, p_1, and p_2 as fixed, requires that

$$p_1 = \frac{P - p_2(dx_2/dQ)}{dx_1/dQ}. \qquad (A.10)$$

Substituting (A.3) into (A.10) and the resulting expression into (A.9), we have

$$\frac{P - p_2\alpha_2 - p_2Q(d\alpha_2/dQ)}{\alpha_1 + Q(d\alpha_1/dQ)}\, x_1 - c(x_1) < PQ - c(x_1) - p_2x_2. \quad (A.11)$$

Cancel $c(x_1)$ from both sides and multiply through by $\alpha_1 + Q(d\alpha_1/dQ)$, which is positive. Then, we may write (A.11) as

$$x_1\!\left(P - \frac{p_2x_2}{Q} - p_2Q\,\frac{d\alpha_2}{dQ}\right) < (PQ - p_2x_2)\left(\frac{x_1}{Q} + Q\,\frac{d\alpha_1}{dQ}\right),$$

since $\alpha_i = x_i/Q$ along the Pareto ray. Now expand both sides,

$$x_1P - \frac{x_1p_2x_2}{Q} - x_1p_2Q\,\frac{d\alpha_2}{dQ} < x_1P - \frac{x_1p_2x_2}{Q} + PQ^2\,\frac{d\alpha_1}{dQ} - p_2x_2Q\,\frac{d\alpha_1}{dQ}$$

Cancel x_1P and $x_1p_2x_2/Q$ from both sides, divide through by Q, and factor $d\alpha_1/dQ$ on the RHS,

$$-x_1p_2(d\alpha_2/dQ) < (PQ - p_2x_2)(d\alpha_1/dQ) \qquad (A.12)$$

Dividing through by $d\alpha_1/dQ$ preserves the direction of the inequality because of our stated condition that $d\alpha_1/dQ > 0$,

$$-x_1p_2\,\frac{d\alpha_2/dQ}{d\alpha_1/dQ} < PQ - p_2x_2,$$

which, from equation (A.4), reduces to

$$-x_1p_2(\alpha_2/\alpha_1) < PQ - p_2x_2$$

Then, since $x_1/\alpha_1 = Q$

$$-Q\alpha_2 p_2 < PQ - p_2 x_2$$

And, since $Q\alpha_2 = x_2$, we have

$$-p_2 x_2 < PQ - p_2 x_2$$

or

$$0 < PQ$$

and the proof is complete.

Notice that this incentive to integrate cannot arise if the final good production function exhibits constant costs. Constant costs imply that $d\alpha_1(Q)/dQ = d\alpha_2(Q)/dQ = 0$. Examination of expression (A.12) then reveals that $\Pi_u = \Pi_I$ in this situation. But with constant costs throughout at the firm level, departure from long-run equilibrium is impossible.

Propositions A.1 and A.2 demonstrate that the upstream monopolist's profits are higher after vertically integrating. Disequilibrium in the downstream industry causes the upstream monopolist's profits to be lower than when the downstream industry is in long-run equilibrium. As a result, the upstream monopolist has an incentive to vertically integrate forward to restore long-run equilibrium. Since this forward integration will continue until the downstream industry has reached a position of long-run equilibrium, the increased profit contains none of the short-run economic profits that emanate from the disequilibrium situation itself.

PART II

Legal Analysis of Vertical Control

7

Legal Treatment of Ownership Integration

The earlier chapters have explored many of the motivations for and consequences of vertical ownership integration and its contractual alternatives. Of course, the main (sole?) motivation is the belief that vertical integration will lead to higher profits. We have explored two avenues to higher profits: (1) the reduction of costs permitted by vertical integration and (2) the enhanced exploitation of existing market power. The first alternative is one that should be applauded, as it clearly leads to increases in consumer welfare. Provided that the existing market power is legal, the second option may or may not be of antitrust concern depending upon the effect on social welfare. If the existing market power is not legal, then that fact should be challenged directly and treated as the horizontal issue that it is. We did not find much support for the notion that vertical integration could be used to *acquire* market power. In interpreting the antitrust laws, however, the courts have raised several objections to vertical integration based upon this possibility. The major objection involves vertical market foreclosure which allegedly occurs when a supplier acquires one of its customers because the supplier's rivals are foreclosed from competing for the acquired firm's business. This objection, we will see, is spurious. Nonetheless, the courts persist in employing the notion of vertical market foreclosure when deciding vertical control cases. In addition, this idea is also popular with the antitrust authorities. We shall see that the Department of Justice Merger Guidelines have used the concept explicitly. Such use is misguided.

Vertical Integration and the Sherman Act

Robert Bork[1] has argued convincingly that the courts have been hostile toward vertical integration and vertical mergers[2] almost from the inception of antitrust. There were five early cases decided between 1911 and 1920 that establish the initial judicial hostility toward vertical integration. We shall be concerned with only two of these, since they contain the genesis of the judiciary's attitude of distrust toward vertical integration. In the *American Tobacco* decision [*United States* v. *American Tobacco Co.,* 164 Fed. 700 (S.D.N.Y., 1908), rev'd 221 U.S. 106 (1911), the first Sherman Act case to deal with vertical integration], one can find an early concern with vertical market foreclosure. American Tobacco had acquired the dominant producer of licorice paste, which is used in producing plug tobacco. The Court was concerned that no rival could survive in the production of plug tobacco because it would have to acquire its supplies of licorice paste from American Tobacco. It expressed a similar concern with respect to the potential for new entry. Thus, the Court seemed to think that vertical integration could create monopoly power by denying competitors access to essential inputs. This, of course, is only possible if a new entrant can neither acquire the essential input elsewhere nor produce it itself.

The *Corn Products* case [*United States* v. *Corn Products Refining Co.,* 234 Fed. 964 (S.D.N.Y., 1916)] introduced a few new wrinkles. Corn Products had some monopoly power in the glucose market, but it was constantly being eroded by entry. For some reason, Corn Products tried to expand control downstream into the syrup market. It drove independent syrup producers out of the industry by engaging in a price squeeze. Corn Products kept the syrup and glucose prices so close together that the independents could not survive.

In addition, Corn Products allegedly engaged in the predatory pricing of glucose. This was supposed to be financed by profits earned on end products. We have here the vertical analog of the usual story of predatory pricing.

Thus, the *American Tobacco* and *Corn Products* decisions reveal a judicial concern with market foreclosure, erection of entry barriers, price

---

[1] This section relies heavily upon the authoritative article by Robert Bork (1954). The Sherman Act deals with vertical integration through its prohibitions of restraints of trade (§1) and of monopolization (§2), 15 U.S.C.A. §§1–2 (1980).

[2] Mergers and acquisitions are related transactions. In a merger, two or more firms are joined together to form a new business entity. All of the parties lose their earlier identity. In an acquisition, the acquired firm disappears while the acquiring firm retains its identity. We shall follow convention and refer to mergers and acquisitions as mergers.

squeezes, and predatory pricing. Except for the predatory pricing, which has a rich history independent of vertical integration,[3] these concerns have carried forward to the last group of cases tried under the Sherman Act. For example, the *Yellow Cab*[4] decision reflected the Court's concern with foreclosure possibilities.

The Checker Cab Manufacturing Co. acquired control of the Yellow Cab Co. and a few other taxi cab companies. Following the acquisitions, it required these subsidiaries to purchase their cabs from Checker. The Justice Department filed suit under the Sherman Act on the theory that Checker's actions illegally foreclosed a substantial fraction of the market for taxi cabs. The lower court observed quite sensibly that such foreclosure was commonplace in vertically integrated firms. The Supreme Court, however, felt otherwise: "By excluding all cab manufacturers other than [Checker] from that part of the market represented by the cab operating companies under their control, the appellees effectively limit the outlets through which cabs may be sold in interstate commerce" (*Yellow Cab* decision, 1947, p. 226). This is a clear endorsement of the market foreclosure doctrine.

The notion of a price squeeze was contained in the *A&P* case [*United States* v. *New York Great Atlantic & Pacific Tea Co.*, 67 F.Supp. 626 (E.D. Ill. 1946), affirmed 173 F.2d 79(7th Cir, 1949)]. For many years, A&P had used its power as a large buyer to get preferential prices from its suppliers. It was partially vertically integrated. To ensure compliance with its demands, A&P threatened further vertical integration. In addition, A&P set up a subsidiary, the Atlantic Commission Co., to buy fruits and vegetables for A&P and for sale to some of A&P's rivals. The court saw the opportunity for a price squeeze here whereby the subsidiary would provide lower prices and free services to A&P, while A&P's rivals would have to pay for the services.

The potential for creating monopoly power through vertical integration is clearly stated in the *Paramount Pictures* case [*United States* v. *Paramount Pictures*, 334 U.S. 131 (1948)]. In this case, the government argued that vertically integrating the production, distribution, and exhibition of motion pictures constitutes a per se violation of the Sherman Act. The

---

[3] The predatory-pricing literature falls into two main groups. The first group examines empirical evidence of alleged instances of predatory pricing. John McGee (1980) provided references to his seminal contribution and to subsequent work. Predatory pricing receives scant empirical support. The second group consists of a complicated squabble over just what should make a pricing decision predatory. This was led by Areeda and Turner (1975). McGee (1980) provided a critique of this literature along with the appropriate references.

[4] *United States* v. *Yellow Cab Co.*, 322 U.S. 218 (1947). In addition to Bork (1954), an interesting account of this case is provided by Kitch (1972).

Supreme Court, however, rejected a per se treatment of vertical integration. Instead, it held that vertical integration violates the Sherman Act when it is "a calculated scheme to gain control over an appreciable segment of the market and to restrain or suppress competition, rather than an expansion to meet legitimate business needs." In addition, a vertically integrated firm will constitute an illegal monopoly if it has the power to exclude competitors, coupled with a purpose or intent to do so.

Whether a particular instance of vertical integration created monopoly power was found to depend upon "the nature of the market to be served . . . and the leverage on the market which the particular vertical integration creates or makes possible" (*Paramount Pictures* decision, 1948, p. 174).

Vertical Integration and the Clayton Act

Prior to 1950, the government had to deal with vertical integration and vertical mergers under the Sherman Act. From 1914 until 1950, Section 7 of the Clayton Act dealt only with horizontal mergers. In 1950, however, Section 7 was amended to extend its prohibition to vertical and conglomerate mergers as well.[5] The amended Section 7 prohibits a vertical merger whenever its effect may be substantially to lessen competition or tend to create a monopoly. A review of the legislative history of the amendment sheds no light on the standards that the courts were expected to use in evaluating vertical mergers. The statute is similarly unenlightening. Thus, it was left to the courts to interpret the statute as best they could. Unfortunately, this exercise has not been terribly successful.

The first case decided by the Supreme Court after Section 7 was amended involved du Pont's acquisition of General Motors stock.[6] In the 1917–1919 period, du Pont bought a 23% stock interest in General Motors. For the Court, the primary issue was "whether du Pont's commanding position as General Motors' supplier of automotive finishes and fabrics was achieved on competitive merit alone, or because its acquisition of the General Motors' stock . . . led to insulation of most of the General Motors' from free competition." Thus, it is clear that the Court was

---

[5] The current Section 7 prohibits any merger or acquisition "where in any line of commerce in any section of the country, the effect of such acquisition may be substantially to lessen competition, or to tend to create a monopoly."

[6] *United States* v. *E.I. du Pont de Nemours & Co.* (*General Motors*), 353 U.S. 586 (1957). This case was actually filed in 1949 before Section 7 was amended. The decision was rendered in 1957 and seems to have been influenced by the amendment.

concerned with market foreclosure. After defining the relevant market to be automotive finishes and fabrics, it followed immediately that General Motors' share was substantial. Accordingly, the stock acquisition resulted in the foreclosure of competition in a substantial share of the market. Indeed, du Pont supplied 67% of General Motors' requirements for finishes in 1946 and 68% in 1947. For fabrics, du Pont supplied 52.3% of requirements in 1946 and 38.5% in 1947.

Given the product market definition and General Motors' dominance in the automobile industry, it is not surprising that the Supreme Court found that "the bulk of du Pont's production has always supplied the largest part of the requirements of the one customer in the automobile industry connected to du Pont by a stock interest. The inference is overwhelming that du Pont's commanding position was promoted by its stock interest and was not gained solely on competitive merit." Accordingly, the Court ruled against du Pont. Some General Motors stockholders suspected that General Motors had been injured by the illegal relationship between damages, but they did not prevail on the merits. [See *Gottesman* v. *General Motors Corporation*, 436 F.2d 1205 (2d Cir,), certiorari denied, 403 U.S. 911 (1971).]

The first case that was filed and decided under the amended Section 7 was the *Brown Shoe* case [*Brown Shoe Company* v. *United States,* 370 U.S. 294 (1962)]. At the time, Brown was a vertically integrated firm that manufactured and retailed shoes. In 1955, Brown was the fourth largest shoe manufacturer in the country with about 4% of the country's total output. In addition, Brown owned, operated or controlled over 1230 retail outlets. As part of its effort to expand at the retail level, Brown acquired Kinney, which was primarily a retailer of shoes.[7] At the time of the trial, Kinney was operating over 400 stores in more than 270 cities. Kinney's share of the total retail sales was about 1.6%. At the time of the merger, Kinney did not buy any shoes from Brown, but by 1957 Brown had become the largest outside supplier of Kinney's shoes. It supplied 7.9% of all Kinney's requirements.

The Court expressed a concern that a vertical merger can foreclose the competitors of either party from a segment of the market otherwise open to them. As a result, it may act as a "clog on competition" and thereby deprive rivals of a fair chance to compete. As this pertains to the *Brown Shoe* facts, the Court provided no economically sensible reason for pro-

---

[7] Kinney also owned four shoe factories that provided about 20 percent of its requirements. At the production level, the Court ruled that the horizontal effects of the merger were insufficient to rule against the merger.

hibiting the merger. It spoke of a vertical merger trend in the industry, but this trend could not hurt competition. Within the Court's own logic there was room for about 100 shoe manufacturers that were fully integrated. See Bork (1978, p. 211). Nonetheless, the merger was banned to prevent market foreclosure in the 1–2% range.

A later case involved Ford's acquisition of a spark plug manufacturer [*Ford Motor Company* v. *United States,* 405 U.S. 562 (1972)]. Although Ford made many of the parts that go into an automobile, it did not make spark plugs. Rather than enter the spark plug industry de novo, Ford acquired the Autolite trade name and a spark plug plant from the Electric Autolite Co. At the time of the acquisition, General Motors made the AC brand for its cars, Champion supplied Ford, Autolite supplied Chrysler, and American Motors bought from Champion and Autolite. After the acquisition, Champion's share of total output declined and market shares were rearranged. The Court objected to the merger on two grounds. First, Ford was poised at the edge of the spark plug market ready to enter de novo. Thus, Ford was a potential entrant and thereby served a desirable function in tempering the pricing freedom of the spark plug manufacturers. But if Ford had entered de novo, that would have been legal. The Court seems not to understand that the results would have been substantially the same as those that followed the merger.

Second, the Court was concerned that the acquisition foreclosed Ford's share of the market from competitors of Autolite. But if market foreclosure has any adverse effects on competition, it must lessen the competition of rivals. In this instance, however, the Court thought that Ford was going to foreclose its purchases in order to disadvantage other spark plug manufacturers. Since that action would not have benefitted Ford, it would not have made much sense.

In their multi-volume treatise, Areeda and Turner (1980, IV, pp. 296–319) surveyed 43 vertical merger cases. In case after case, vertical market foreclosure is a primary concern of the court involved. None of these cases provides a sound economic rationale for the judiciary's fear of foreclosure. Apparently, the courts have adopted a presumptive approach to vertical mergers. If either market is highly concentrated and the merger results in foreclosure which is not *de minimus,* then the merger is apt to be held unlawful. Using this approach greatly simplifies the court's burden—it just rules according to the presumption without having to deal with an economic analysis of the facts. Lawrence Sullivan (1977, p. 663) remarked "if the share foreclosed is high enough, the prosecutor need not articulate any particular theory about why the merger is injurious; it need only insist that it is."

Department of Justice Merger Guidelines, 1968

In 1968, the Antitrust Division of the Department of Justice issued merger guidelines to alert the business community and the legal profession to the standards that would be used to determine whether a specific merger would be challenged. [See Department of Justice Release, May 30, 1968; reprinted, 360 ATRR pp. x-1 et seg. (June 4, 1968).] Of course, the reason for setting out these standards was to deter questionable mergers. This would make actual enforcement through litigation unnecessary. The 1968 guidelines were quite hostile toward mergers generally and toward vertical mergers in particular. This attitude reflected the prevailing antitrust view of vertical integration.

All types of mergers—horizontal, vertical, and conglomerate—were covered by the 1968 Merger Guidelines. With respect to vertical mergers, the Department of Justice enforcement activity was intended to prevent changes in market structure that eventually would have significant anti-competitive consequences. Generally, the Department of Justice believed that such consequences would be expected to occur whenever a particular vertical merger in a supplying or purchasing market tended significantly to raise barriers to entry in either market or to disadvantage existing nonintegrated or partially integrated firms in either market in ways that were unrelated to economic efficiency.

The Department of Justice recognized that barriers to entry stemming from economies of scale in production and distribution are not avoidable without significant penalties. It was alleged, however, that vertical mergers tend to raise barriers to entry in undesirable ways: (1) by foreclosing equal access to potential customers, thereby reducing the ability of nonintegrated firms to capture competitively the market share needed to achieve an efficient level of production, or imposing the burden of entry on an integrated basis (i.e., at both the supplying and purchasing levels) even though entry at a single level would permit efficient operation; (2) by foreclosing equal access to potential suppliers, thereby either increasing the risk of a price or supply squeeze on the new entrant or imposing the additional burden of entry as an integrated firm; or (3) by facilitating promotional product differentiation, when the merger involves a manufacturing firm's acquisition of firms at the retail level. The Department of Justice was also concerned that in addition to impeding the entry of new sellers, the foregoing consequences of vertical mergers would also constrain the expansion of existing sellers by conferring on the merged firm competitive advantages, unrelated to real economies of production or distribution, over nonintegrated or partly integrated firms. While the department conceded that in some cases vertical integration may raise barriers to entry or disadvantage existing competitors only as the result of the achievement of significant economies of production or distribution, it was nevertheless convinced that integration accomplished by a large vertical merger will usually raise entry barriers or disadvantage competitors to an extent not accounted for by, and wholly disproportionate to, such economies as may result from the merger.

Supplying Firm's Market

In determining whether to challenge a vertical merger on the ground that it may significantly lessen existing or potential competition in the supplying firm's market, the Department of Justice attached primary significance to (1) the market share of the supplying firm, (2) the market share of the purchasing firm or firms, and (3) the conditions of entry in the purchasing firm's market. Accordingly, the Department of Justice would ordinarily challenge a merger or series of mergers between a supplying firm, accounting for approximately 10% or more of the sales in its market, and one or more purchasing firms, accounting in toto for approximately 6% or more of the total purchases in that market, unless it clearly appears that there is no significant barrier to entry into the business of the purchasing firm or firms.

Purchasing Firm's Market

Although the standard described in the preceding paragraph was designed to identify vertical mergers having likely anticompetitive effects in the supplying firm's market, adherence by the Department of Justice to that standard will also normally result in challenges being made to most of the vertical mergers that may have adverse effects in the purchasing firm's market. This follows because adverse effects in the purchasing firm's market will normally occur only as the result of significant vertical mergers involving supplying firms with market shares in excess of 10%. There are, however, some important situations in which vertical mergers that would not be subject to challenge under the preceding paragraph would nonetheless be challenged by the Department of Justice on the ground that they raise entry barriers in the purchasing firm's market, or disadvantage the purchasing firm's competitors by conferring upon the purchasing firm a significant supply advantage over unintegrated or partially integrated competitors or potential competitors.

Thus, in the absence of any clearly articulated economic theory, the Department of Justice adopted a hostile, structural approach to dealing with vertical mergers. Much of the concern surrounding vertical mergers focuses on the notion of market foreclosure. We shall turn to a critique of the market foreclosure doctrine.

Market Foreclosure Doctrine

As we have seen previously, the legal analysis of vertical integration has focused primarily on the concept of market foreclosure. Consequently, we should examine this doctrine in some detail. The basic idea behind the foreclosure doctrine is that an input supplier, by merging with one of its customers, effectively removes that firm's purchases from the open market.[8] By so doing, vertical integration reduces the size of the market that is available to the acquiring firm's nonintegrated rivals in the industry.

Although discussions of the evils of foreclosure are notable in their level of obfuscation, we may infer from them two anticompetitive effects that are alleged to result. First, by decreasing the size of the "open" market, a vertical merger may increase the level of concentration in the nonintegrated segment of the market. For example, suppose we have three firms at the upstream stage, each with one-third of the intermediate-product market. If one of these firms merges with all of its downstream customers, the remaining two firms that have not vertically integrated will now share the "open" market equally. The single-firm concentration ratio in the "open" market has increased from one-third to one-half as a result of the foreclosure that results from vertical merger. If the firm that vertically integrates enjoys a relatively large share of the market prior to the merger, it is quite possible that concentration in the "open" segment of the market will fall as a result of foreclosure. The second anticompetitive effect that is alleged to occur is that, by foreclosing the integrated segment of the market to nonintegrated firms but not foreclosing access to the nonintegrated segment of the market to the integrated firms, vertical mergers place the nonintegrated firms at a competitive disadvantage. This, in turn, is expected to lead to exit by these firms over time, thereby increasing the level of concentration in the overall market as well.

The foreclosure doctrine has not been generally well received by economists. This cool reception is due to the fact that the fundamental presumption underlying this doctrine—that vertical merger entirely removes a given set of transactions from the pressure of market forces—does not make economic sense. The basic flaw in the legal analysis of vertical

---

[8] Of course, an analogous argument can be made when an input customer merges with one of its suppliers.

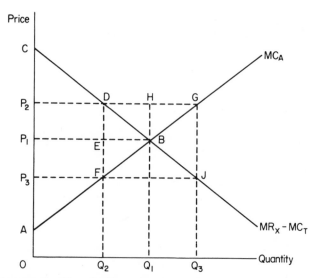

Figure 7.1 Integrated-firm equilibrium and the foreclosure doctrine. From Allen, Bruce T. "Vertical Integration and Market Foreclosure: The Case of Cement and Concrete," *Journal of Law and Economics*, Vol. 14 (April 1971), pp. 251–274. © by The University of Chicago. All rights reserved.

integration was exposed most clearly by Bruce Allen (1971)[9]. Allen's analysis is repeated here with the aid of Figure 7.1.

This figure pertains to the situation faced by a vertically integrated firm that manufactures an input A that is employed by the firm's downstream division in the production of output X. The curve labeled MC_A represents the marginal cost of producing input A. The curve labeled $MR_X - MC_T$ is the downstream division's derived demand for the input,[10] which is given by the marginal revenue of output MR_X minus the marginal cost of transforming the input into output, MC_T. This curve may slope downward because of a downward sloping demand for the final product, an upward sloping MC_T curve, or both.

Now, suppose that there are nonintegrated firms at both stages of production so that an active intermediate-product market for input A exists. Then, will the fact that this firm is vertically integrated remove its operations from the influences of this market? Of course, the answer is no. The vertically integrated firm will always find it optimal to set the transfer price of the intermediate good equal to the market price of the input. It will then participate in the intermediate-product market if the market

[9] The earlier articles by Hirschleifer (1956) and by Liebeler (1968) are also helpful in understanding the basic problem with this doctrine.

[10] See Chapter 3 for the derived demand model.

price differs from P_1 in the graph, selling input A if its price exceeds P_1 and buying it if its price falls short of P_1.

To see this, first suppose that the open-market price is equal to P_1. Then, to maximize profits the upstream division will transfer Q_1 units of the input to the downstream division at the market price P_1. In this case, the integrated firm does not participate in the intermediate-product market at all. It manufactures its entire requirements internally and utilizes its entire output of the intermediate good. Total profits to the downstream division are given by the triangle P_1BC (the area under the marginal revenue product curve and above the transfer price), and total profits to the upstream division are P_1BA (the area above the marginal cost curve and below the transfer price). Consequently, total profits to the overall integrated operation are ABC.

Given the market price of P_1, any other combination of transfer price and input quantity will result in less (or no more) profits being earned by the integrated firm. For example, suppose that the upstream division forces the downstream division to pay a price equal to P_2, despite a market price of P_1. If the downstream division is not coerced into buying more, it will purchase Q_2 units of the input and earn a profit of P_2DC, thereby losing P_1BDP_2 as compared to what it earned before the upstream division raised its price above the market level. From this loss, the upstream division will gain back only P_1EDP_2, leaving a total reduction in profit on sales to the downstream division of FBD. Part of this loss may be offset if the upstream division sells $Q_1 - Q_2$ units on the open market at the market price of P_1. But this still leaves a loss of EBD that is not offset.

Now, suppose that the upstream division forces the downstream division to purchase Q_3 units of the input at the price P_2. This increases the profits of the upstream division by the area P_1BGP_2. But this entire increase is paid by the downstream division. In addition, production costs are increased so that total profits are once again reduced. With an open market price of P_1, the best that the integrated firm can do if it insists on charging a transfer price of P_2 is to force the downstream division to purchase Q_1 units of the input (the same amount it would have purchased willingly at the market price P_1). The downstream division then loses P_1BHP_2 and the upstream division gains P_1BHP_2 as compared to the situation in which the transfer price is set at the market level. Thus, total profits are the same. The only thing that is accomplished in this case is a redistribution of profits from the downstream to the upstream stage.

What happens if the open market price of the intermediate product is not equal to P_1? For instance, suppose this price rises to P_2. By the same line of reasoning used above, the vertically integrated firm facing an open market price of P_2 will set the internal transfer price at P_2. Then, the downstream division will purchase Q_2 units of the input and the upstream division will sell an additional $Q_3 - Q_2$ units on the open market. Profits to the downstream division are P_2DC, and profits to the upstream division are AGP_2. No other solution can yield greater total profits to the integrated operation. Conversely, if the open market price falls below P_1, say to P_3, then the transfer price will be reduced to that lower level. Here, the downstream division will purchase Q_2 units from the upstream division and an additional $Q_3 - Q_2$ units on the open market. Profits to the upstream division are AFP_3 and profits to the downstream division are P_3JC.

In summary, we conclude that the profit-maximizing vertically integrated firm will always set the transfer price of the internally supplied input at the level determined by the open market. This firm will continue to participate in that market (either as a buyer or a seller) as long as the market price differs from the level at which the internal supply and demand curves intersect. It follows from this that the foreclosure doctrine has no foundation in microeconomic theory.

This theoretical reasoning can be applied to the Brown Shoe foreclosure. Brown was initially in a position where it transferred some of its shoe production internally and sold some of it to downstream competitors. In Figure 7.2, assume that the market price was P_1 and that Brown's total production was Q_1. It transferred Q_2 to its retail outlets and sold $Q_1 - Q_2$ units to independent retailers. When Brown acquired Kinney, the

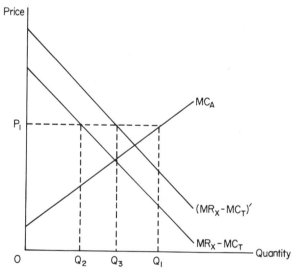

Figure 7.2   Effect of an acquisition on the integrated firm's equilibrium.

internal demand shifted to $(MR_X - MC_T)'$. Thus, Brown began to supply Kinney with its shoes. Now, Brown transferred Q_2 units to its original retail outlets and $Q_3 - Q_2$ to the newly acquired Kinney outlets. Only $Q_1 - Q_3$ remains to be sold to outsiders. It is clear that to a certain extent Brown displaced some of Kinney's former suppliers. But some of Brown's former customers were also displaced. There is no apparent reason why the displaced suppliers and the displaced customers cannot get together. Accordingly, the foreclosure really does not impose any hardships on Brown's rival producers or on Brown-Kinney's rival retailers.

The only way that Brown's acquisition of Kinney could hurt Kinney's former suppliers is if Brown has excess capacity. In that case, Brown could continue to supply all of its former customers and supply some of Kinney's needs. As a result, the former suppliers lose business permanently. But this is not necessarily bad. If there is excess capacity in an industry, some of that capacity should exit the industry. It should be apparent that the least efficient producers will leave. It is not in Brown's interest to preserve its own inefficient productive capacity if it must sell them at retail because this inefficiency will lead to lower profits or losses at the retail stage. The losses may be reported elsewhere, but the problem is at the production stage.

The 1982 Merger Guidelines

There is no evidence that the courts have recognized the fallacies in the market foreclosure doctrine. But the Department of Justice has seen the light—at least temporarily. President Reagan's appointment of William Baxter as Assistant Attorney General in charge of the Antitrust Division has led to a substantially different attitude toward many business practices. In an effort to reduce uncertainty surrounding the altered attitudes toward mergers, the Department of Justice issued revised guidelines in 1982. [See Department of Justice release, Merger Guidelines, June 4, 1982; reprinted, 1069 ATRR pp. S-1–S-16 (June 17, 1982).] These new guidelines reflect a more enlightened attitude toward vertical integration via the merger route. Accordingly, the focus of the new guidelines is squarely on horizontal market power. Vertical and conglomerate mergers are treated as "nonhorizontal" mergers that may have some horizontal effects. Three competitive problems resulting from vertical mergers were identified as (1) increased entry barriers, (2) likely facilitation of collusion, and (3) possible evasion of price regulation.

Barriers to Entry

The Department of Justice contends that under certain conditions vertical mergers could create anticompetitive entry barriers. There are, however, three necessary but not sufficient conditions for this undesirable consequence. (See Chapter 3 for a more thorough treatment of the potential relationship between vertical integration and entry barriers.) First, the degree of vertical integration between the two markets must be so extensive that entrants to the "primary market" also would have to enter the "secondary market" simultaneously. Second, the requirement of entry at the secondary level must make entry at the primary level significantly more difficult and less likely to occur. Finally, the structure of the primary market must be otherwise so conducive to noncompetitive performance that the increased difficulty of entry is likely to affect its performance.

If there is sufficient unintegrated capacity in the secondary market, new entrants to the primary market would not have to enter both markets simultaneously. The Department of Justice is unlikely to challenge a merger on this ground where postmerger sales (purchases) by unintegrated firms in the secondary market would be sufficient to service two minimum-efficient-scale plants in the primary market. Simultaneous entry, however, acquires competitive significance only in that it is substantially more difficult than entry at the primary stage alone. If entry at the secondary level is easy in absolute terms, the requirement of simultaneous entry to that market is unlikely to affect adversely entry to the primary market. Finally, entry barriers are unlikely to affect performance if the structure of the primary market is otherwise not conducive to monopolization or collusion. The Department of Justice is unlikely to challenge a merger on this ground unless overall concentration in the primary market generates a Herfindahl Index above 1800.

Facilitating Collusion

The Department of Justice recognized that retail prices are generally more visible than prices in upstream markets. As a result, vertical integration by upstream firms may facilitate collusion by making it easier to monitor prices. Obviously, this cannot pose a serious problem unless the upstream industry's structure is conducive to collusion. Accordingly, the Department of Justice is unlikely to challenge a merger on these grounds unless the Herfindahl Index exceeds 1800 at the upstream stage. In addition, the extent of vertical integration must be substantial so that a large percentage of the upstream product will be sold through vertically integrated outlets.

Vertical integration may also facilitate collusion if a disruptive buyer is eliminated. (A disruptive buyer is one that is a sufficiently attractive customer so that the upstream firms cannot resist competing for its business.) If one of the suppliers buys that customer, then collusion at the upstream stage may be easier to maintain. This, of course, depends upon the upstream industry's structure being conducive to collusion. Accordingly, the Department of Justice is not likely to challenge a vertical merger on these grounds, unless the Herfindahl Index exceeds 1800. Moreover, the so-called disruptive buyer must be substantially larger than other customers.

Evasion of Rate Regulation

It is well known that a regulated public utility can acquire an input supplier and evade regulation (Chapter 6). It simply pays its input subsidiary a noncompetitive price for the input and passes the overcharge along to its customers. In effect, it transfers profit from the regulated industry to an unregulated industry. While this is not usually an antitrust problem, the Department of Justice is concerned about consumer welfare and will consider challenging mergers that create substantial opportunities for such abuses.

Concluding Remarks

Economists do not favor the market foreclosure doctrine. Moreover, we have seen a host of alternative and more logically appealing explanations of observed vertical integration. Nonetheless, the antitrust authorities and the courts have consistently relied on the market foreclosure doctrine in dealing with vertical mergers. This fundamental inconsistency between the legal and the economic approaches to vertical integration has persisted for a long time.

The virtual fixation of enforcement officials on the foreclosure doctrine as the appropriate model with which to judge the competitive effects of vertical integration can be seen in the court decisions and the past merger guidelines. The current guidelines indicate that the Antitrust Division has revised its attitude and does not focus on foreclosure. This is a most welcome step in the right direction.

8

Per Se Illegal Contractual Controls

The legal status of the contractual alternatives to vertical ownership integration varies from per se illegality to presumptive legality. In this chapter, we shall examine three contractual alternatives that suffer per se illegality: maximum-resale price-fixing, resale price maintenance, and tying arrangements. Per se illegality makes employing these methods of vertical control extremely hazardous. A plaintiff's burden in such situations is greatly reduced because the court will infer the adverse economic effect from the mere existence of the business practice. Since these three business practices are hard to disguise, the firm employing such devices is vulnerable to legal proceedings from several directions. First, the antitrust authorities can charge the firm and the individuals involved with violating the Sherman Act, the Clayton Act, or the Federal Trade Commission Act. In the case of Sherman Act prosecution, the potential penalties can be quite severe. The firm can be fined as much as $1,000,000. For the individuals involved, the sanctions could extend to a 3-year jail sentence and/or a $100,000 fine. Second, the firm can be sued for damages caused by the tying arrangements. The authority for damage suits is Section 4 of the Clayton Act, which provides "that any person who shall be injured in his business or property by reason of anything forbidden in the antitrust laws may sue therefore . . . and shall recover threefold the damages by him sustained (15 U.S.C.A. 15).

In addition to sketching the legal development pertaining to each contractual alternative, we shall offer a brief critique of the legal doctrine.

Fixing Maximum-Resale Prices

The legal treatment of fixing maximum-resale prices is short, but not particularly sweet for the consumer. The earliest mention of maximum prices can be found in the Supreme Court's decision in *United States* v. *Socony-Vacuum Oil Co.* Although this case did not deal with fixing maximum resale prices, Justice Douglas wrote that "under the Sherman Act a combination formed for the purpose and with the effect of raising, depressing, fixing, pegging, or stabilizing the price of commodity in interstate or foreign commerce is illegal *per se*" [310 U.S. 150, 223 (1940)]. For over 10 years, the language regarding "depressing" prices remained nothing more than dictum. In 1951, however, the Supreme Court handed down an important decision in *Kiefer-Stewart Co.* v. *Joseph E. Seagram & Sons* [340 U.S. 211 (1951)].

Kiefer-Stewart was an Indiana firm that had a wholesale liquor business. Seagram and Calvert were producers of liquor that was sold to wholesalers in Indiana. Seagram and Calvert agreed not to sell their products to any wholesaler that refused to respect *maximum*-resale prices set by Seagram and Calvert. Since Kiefer-Stewart refused to respect these maximum-resale prices, it was denied access to Seagram and Calvert products. As a result, Kiefer-Stewart was injured due to lost sales. Seagram and Calvert claimed that the decision to fix maximum-resale prices was motivated by a horizontal price-fixing conspiracy among its wholesale customers and presented evidence to support this contention. In spite of this evidence, the Court ruled in favor of Kiefer-Stewart. It explicitly affirmed the dictum in *Socony-Vacuum* on the grounds that agreements to fix maximum-resale prices "cripple the freedom of traders and thereby restrain their ability to sell in accordance with their own judgment" (*Kiefer-Stewart*, p. 213).

Following the logic developed in *Kiefer-Stewart,* the Court held a maximum price-fixing scheme to be illegal in *Albrecht* v. *Herald Co* [390 U.S. 145 (1968)]. Due to the special nature of the home delivery of newspapers, the publisher of the *Globe-Democrat,* a St. Louis newspaper, assigned exclusive territories to its carriers. Within each territory the assigned distributor had a monopoly on home delivery. This practice assured that the costs of providing home delivery service would be minimized because duplicate effort was eliminated. These delivery routes were subject to termination if the carrier charged a price in excess of the price advertised by the *Globe-Democrat.* Albrecht was one of the *Globe-Democrat's* distributors. Although he was aware of the maximum-price policy, he ignored the policy and charged a higher price. Following the complaints of several customers, the publisher warned Albrecht that he was jeopardizing his distributorship. When Albrecht continued to overcharge his customers, the publisher took action against him by first competing directly and later by substituting another distributor for part of the territory. When Albrecht filed suit he was terminated and forced to sell his distributorship.

He sued the publisher and others for injuries suffered. The Supreme Court explicitly approved its earlier decision in *Kiefer-Stewart:* "We think *Kiefer-Stewart* was correctly decided and we adhere to it" (*Kiefer-Stewart*, p. 152). This, of course, doomed the Herald Co. as the Court found that "schemes to fix maximum prices, by substituting the perhaps erroneous judgment of a seller for the forces of the competitive market, may severely intrude upon the ability of buyers to compete and survive in the market" (*Kiefer-Stewart*, p. 152). In spite of the fact that the Court was aware that competitive forces may not operate in exclusive territories, it ruled that fixing maximum prices violated Section I of the Sherman Act.

The rule against fixing maximum prices persists today. Although the practice may occur in any situation where successive market power exists, most of the recent cases involve newspaper distribution[1] or gasoline dealers.[2] In these cases, the lower courts have heartily endorsed the unambiguous holding of the Supreme Court. Consequently, the vitality of the rule enunciated in *Albrecht* continues despite its questionable economic logic.

Maximum-resale price-fixing is invariably used by a supplier to prevent its distributors from exploiting whatever market power they possess. In *Kiefer-Stewart,* ostensible horizontal competitors in wholesale distribution were conspiring to *raise* prices. The actions of Seagram and Calvert can be seen as a means of preventing the inevitable decline in sales that accompanies an increase in price. Although each wholesale distributor may have had precious little monopoly power, collectively the wholesalers were trying to emulate the price and output that a monopolist would select. Fixing maximum prices is a way of thwarting these intentions of horizontal price-fixers.

The analysis of *Albrecht* is even more direct. The distributor, Albrecht, had an exclusive territory. For home delivery, Albrecht had a complete monopoly in his territory. The publisher wanted to prevent Albrecht from behaving like the monopolist that he was. Fixing the maximum-resale price of the *Globe-Democrat* was the publisher's way of doing that.

In both of the cases afforded full Supreme Court review, we see that the manufacturer enjoyed some market power through product differentiation (brand names of Seagram and Calvert) or through being the only seller (of an evening newspaper). The distributor also enjoyed some market power through the collusion of horizontal competitors (*Kiefer-Stewart*) or

---

[1] See, e.g., *Knutson* v. *The Daily Mirror,* No. C-73-1354-CBR, ND. Cal. (1979) and *Auburn News Company, Inc.* v. *Providence Journal Company,* No. 80-0446, D.R.I. (1980).

[2] See, e.g., *American Oil Company* v. *Arnott,* No. 79-1357 (1980) and *Yentsch* v. *Texaco, Inc.,* No. 79-7735 and 79-7746 (1980).

through exclusive territories (*Albrecht*). By preventing maximum-resale price-fixing, the antitrust law promotes the welfare of the local monopolists at the expense of the consumer.

Resale Price Maintenance

While the economic analysis of resale price maintenance may not lead to unambiguous conclusions regarding consumer welfare, the judicial history is quite clear. Since 1911, the rule on resale price maintenance has been quite clear and consistent: such agreements violate Section 1 of the Sherman Act. In the interlude, ways around this harsh rule have been sought. We shall see that two legal gambits, refusals to deal and agency relationships, have not been wholly successful. Moreover, even special legislation has not fared too well.

Dr. Miles Medical Co. v. John D. Park & Sons Co.

The controlling case is *Dr. Miles* [220 U.S. 373 (1911)], which involved the manufacturer of proprietary medicines that were prepared according to secret formulas. Dr. Miles devised an elaborate pricing system that established minimum prices at which sales of its products would be made throughout the entire distribution chain. The case arose because John D. Park, a wholesale drug concern, refused to enter into the restrictive agreements that Dr. Miles required. Instead, John D. Park obtained the medicines for resale at discount prices from other wholesalers who had accepted the restrictions. Dr. Miles sued John D. Park for maliciously interfering with the contract between Dr. Miles and some of his buyers.

The Supreme Court recognized the issue as being whether the restrictive agreements were valid. If they were, Dr. Miles would win; if they were not, John D. Park would prevail. The Court characterized the contractual scheme as "a system of interlocking restrictions" whereby Dr. Miles could control the prices of its products at every stage in the distribution system. As a result, Dr. Miles would ultimately set the amount that the consumer would pay, thereby eliminating all competition between retailers. Since these agreements obviously restrained trade, the Court's task was to determine whether the restraints violated Section 1 of the Sherman Act.

Dr. Miles made two arguments in its defense. First, it contended that it was entitled to extensive control because the medicines were produced according to a secret process. The Court rejected this argument easily:

[This argument] implies that if, for any reason, monopoly of production exists, it carries with it the right to control the entire trade of the produced article, and to prevent any competition that otherwise might arise between wholesale and retail dealers . . . But, because there is monopoly of production, it certainly cannot be said that there is no public interest in maintaining freedom of trade with respect to future sales after the article has been placed on the market and the producer has parted with his title. [*Dr. Miles,* 220 U.S. 373 (1911)]

The second argument was disposed of just as easily. Dr. Miles argued that a manufacturer is entitled to control the prices on all sales of his own products. In effect, Dr. Miles claimed that it could produce and sell or not as it saw fit and, therefore, that it could impose conditions as to the prices at which purchasers may resell the item. The Court reasoned that "because a manufacturer is not bound to make or sell, it does not follow in cases of sales actually made he may impose upon purchases every sort of restriction. Thus, a general restraint upon alienation is ordinarily invalid" (*Dr. Miles,* p. 380).

The rule against resale price maintenance is robust. Bobbs-Merrill [*Bobbs-Merrill Co.* v. *Straus,* 210 U.S. 339 (1908)] had a copyright on a book and claimed a right to fix its resale price. In that case, the Court held that the copyright privilege was not designed to provide such prerogatives. In *Bauer & Cie* v. *O'Donnell,* 229 U.S. 1 (1913), the manufacturer claimed a right to fix resale prices on the basis of a patent. Since a patent provides the right to restrict use and sale was a use, it had the right to restrict resale price. The Court held that when the manufacturer sold a product whose value is in its use, the right to restrict use passes to the purchaser. Thus, neither copyrights nor patents confer authority for resale price maintenance.

In Dr. Miles, the Court did not announce that resale price maintenance was a per se violation of the Sherman Act. The Court's reasoning, however, could have been used to reach such a conclusion:

agreements or combinations between dealers, having for their sole purpose the destruction of competition and the fixing of prices, are injurious to the public interest and void. . . .
The complainant's [Dr. Miles'] plan falls within the principle which condemns contracts of this class. . . . The complainant having sold its product at prices satisfactory to itself, the public is entitled to whatever advantage may be derived from competition in the subsequent traffic. [220 U.S. 373 (1911)]

Subsequent decisions by the Supreme Court have left little doubt that vertical price-fixing is a per se violation of the Sherman Act. Recently, the Court remarked that "[t]he per se illegality of [vertical] price restrictions has been established firmly for many years," [*Continental T.V., Inc.* v. *GTE Sylvania, Inc.,* 433 U.S. 36, 48, note 18 (1977)]. This case did not involve vertical price-fixing, but the Court does not seem disposed to alter its view on resale price maintenance.

Fair Trade Legislation

Some states passed fair trade laws that would permit vertical price-fixing. These laws were not very helpful by themselves because of the Federal Antitrust laws that prohibited resale price maintenance. Eventually, the Miller-Tydings Amendment was passed, which permitted resale price maintenance where state law permitted it. When the Supreme Court ruled that nonsigners to a fair trade agreement could not be bound by the established minimum prices, the Miller-Tydings Amendment became ineffective. In an effort to revive the force of state fair-trade laws, Congress amended the Federal Trade Commission Act by passing the McGuire Act. This extended fair trade to unwilling resellers.

The purpose of this legislation generally has been to protect small merchants who are threatened by mass merchandisers. Many of these small merchants are quite inefficient and cannot survive in competition with large retailers that enjoy the benefits of greater efficiency. This has resulted in a campaign in Congress to pass a federal resale price-maintenance law. The admitted rationale for such legislation was the competitive difficulties encountered by small merchants. This effort failed. Subsequently, the tide of public opinion turned as inflation alarmed consumers. During the Ford administration, the Consumer Goods Pricing Act of 1975 was passed. This repealed the McGuire Act and restored the supremacy of the Sherman Act over state fair-trade laws.

Refusals to Deal

Some manufacturers have attempted to implement a resale price-maintenance policy by refusing to deal with those who fail to honor the minimum prices. The Supreme Court explicitly approved of this behavior in its *Colgate* decision [*United States* v. *Colgate & Co.*, 250 U.S. 300 (1919)]. Through its resale price-maintenance program, Colgate's products sold at uniform retail prices. In contrast to the Dr. Miles program, Colgate did not have elaborate contractual agreements. Colgate allowed each of its dealers to sell at whatever price the dealer felt appropriate. But Colgate would simply not deal with those dealers who deviated from the minimum prices. Since there were no contractual agreements, the Court approved of Colgate's behavior:

> In the absence of any purpose to create or maintain a monopoly, the [Sherman] Act does not restrict the long recognized right of trader or manufacturer engaged in an entirely private business, freely to exercise his own independent discretion as to

parties with whom he will deal; and, of course, he may announce in advance the circumstances under which he will refuse to sell. [250 U.S. 300 (1919)]

The freedom to engage in resale price-maintenance by relying upon *Colgate* has been severely restricted by subsequent decisions. This is not too surprising when one recognizes that the *Colgate* decision is incoherent. If the manufacturer unilaterally decides to impose resale price maintenance through explicit contracts, then it violates the law according to *Dr. Miles*. If it achieves the same end without explicit contracts, it may be safe on the *Colgate* logic. This is one of the law's infamous distinctions without a difference.

For some reason, the Supreme Court is reluctant to overrule *Colgate*. Instead, it has narrowed the scope of the rule so severely that a manufacturer should be cautioned that "the line between legal and illegal conduct here is a very narrow one and if a seller chooses to walk that line, he must do so at his peril" [FTC Advisory Opinion Digest No. 163 (1968), 16 C.F.R. paragraph 15.163 (1973)]. Currently, a manufacturer can announce a policy of not dealing with those who do not follow his schedule of minimum prices and actually follow through on this policy. If this results in resale price maintenance, the seller is secure under *Colgate*. If, however, the seller takes any additional steps, he will fall off that "narrow line." Some ways of losing the *Colgate* shield are (1) establishing a policing mechanism to detect violators, (2) reinstating violators who promise to abide by the minimum prices in the future [see *FTC* v. *Beech-Nut Packing Co.*, 257 U.S. 441 (1922)], (3) asking wholesalers to police the retailers [see *United States* v. *Parke, Davis & Co.*, 362 U.S. 29 (1960)], and (4) assuring complying dealers that it will do something about those who do not comply [see *United States* v. *GM Corp.*, 384 U.S. 127 (1966)]. In all of these cases, the seller would have overstepped the bounds of its *Colgate* rights.

Agency Relationships

In his dissenting opinion to the *Dr. Miles* decision, Justice Holmes expressed some concern that the majority has confused form with substance because the Dr. Miles program would have been legal if an agency relationship had been established. An agent is one who acts on behalf of another. In *General Electric*, [*United States* v. *General Electric Co.*, 272 U.S. 476 (1927)] the agency relationship carried the day for General Electric. In that case, there was no serious dispute about whether General Electric was setting the subsequent prices for its light bulbs. GE argued, however, that the bulbs were sold through wholesale and retail agents. The bulbs went to the agents under consignment and remained the property of GE until they were ultimately sold. The agents were given no discretion to deal with the bulbs in any way other than as directed by GE. If the middlemen had been purchasers, the Court ruled, prices could not be fixed. In contrast, if the middlemen were agents, then GE can fix the prices. The Court found that the independent merchants were agents of GE in the consignment sales.

The *General Electric* decision is another one that has not been expressly overruled, but the Supreme Court's decision in *Simpson* v. *Union Oil,* 377 U.S. 13 (1964), leaves little of substance standing from the GE case. Union Oil had what appeared to be a consignment agreement with dealers like Simpson. Union Oil owned its retail outlets and leased them to its dealers. These leases were used to enforce the retail price structure since renewal was not automatic. The court found that the lease agreement was used coercively and appeared to be quite effective in maintaining retail prices. Although the consignment agreement was not much different from the one that General Electric used, the Court seemed to be offended by this escape hatch. Justice Douglas said that ''a consignment, no matter how lawful it might be as a matter of private contract law, must give way before the federal antitrust policy.'' Further, ''when, however, a 'consignment' device is used to cover a vast gasoline distribution system, fixing prices through many retail outlets, the antitrust laws prevent calling the 'consignment' an agency for then the end result of United States v. Socony-Vacuum Oil Co., would be avoided merely by clever manipulation of words, not by differences in substance.'' Thus, the consignment or agency exemption from the dictates of *Dr. Miles* was severely attenuated.

There are two anticompetitive reasons for resale price maintenance. First, it can be used as a policing mechanism by a dealer cartel. Second, it can be used as a policing mechanism by a manufacturer cartel. There is only one procompetitive reason for fixing minimum-resale prices: to ensure that product-specific services are provided by dealers.

But majority does not rule in this case. For one thing, it would not be easy for a retailer cartel to impose a resale price-maintenance agreement on the manufacturer. For another thing, there is a way to distinguish between the procompetitive and anticompetitive motivations. If minimum-resale prices are imposed to assure the provision of commodity-specific services, the result should be an expansion of output. In contrast, if either a dealer cartel or a producer cartel is behind the resale price-maintenance scheme, the industry output will fall. Consequently, the antitrust authorities can determine whether the resale price-maintenance agreement offends the spirit of the antitrust laws. As matters stand now, there is no room for procompetitive resale price maintenance.

Tying Arrangements

Tying arrangements may take many forms, but the essence of each such arrangement is a conditional sale. Specifically, the seller of commodity A (the tying good) agrees to sell it only on the condition that the buyer also purchase a separate commodity B (the tied good). The judiciary's instinct[3] has been that tying is used to lever market power from the tying-good market into the tied-good market. A firm that employs tying arrangements may run afoul of Section 1 of the Sherman Act, Section 3 of the Clayton Act, or Section 5 of the Federal Trade Commission Act. The business practice of tying appears to be presumptively illegal. Generally, if (1) a seller conditions the sale of one good on the purchase of a separate tied good, (2) the supplier has sufficient market power in the tying good to restrain competition in the tied good, and (3) a not-insubstantial volume of commerce is affected, then the seller is guilty of an illegal tie. We shall develop this unfortunate conclusion by reviewing the Supreme Court's rulings on tying arrangements.

Prior to the passage of the Clayton Act in 1914, tying cases had to be brought under the Sherman Act. These efforts proved to be largely unsuccessful. For example, in *Henry* v. *A. B. Dick Co.* [224 U.S. 1 (1912)], A. B. Dick had a patent on a duplicating machine. The users of the patented duplicator were required to use A. B. Dick ink. Henry was charged with contributory infringement of the duplicator patent for making and selling a substitute ink to be used in the A. B. Dick duplicator. In response, Henry argued that the Sherman Act had been violated by this tying arrangement and this fact should offset the A. B. Dick's claim. This defense failed. Partly in response to this decision, Congress included a section in the Clayton Act that prohibited tying.

Section 3 of the Clayton Act, however, provided the foundation for an increasingly hostile attitude toward tying arrangements. Generally, Section 3 of the Clayton Act makes it unlawful for a person to sell or lease a commodity on the condition that the buyer or lessee not buy or lease the goods from a competitor of the seller or lessor where the effect may be to substantially lessen competition or tend to create a monopoly. Since the practical effect of a tying arrangement precludes the use of a competitor's wares, Section 3 prohibits tying.

The early cases filed under Section 3 of the Clayton Act developed the leverage theory of tying.[4] For example, in *United Shoe Machinery Corp.*

---

[3] The word "instinct" is Sullivan's (1977), but it is almost a perfect choice since precious little economic analysis supports the hostility accorded this business practice.

[4] For a devastating critique of the leverage theory of tying, see Bowman (1957).

v. *United States*, [258 U.S. 451 (1922)], the owner of a patented commodity leased the patented commodity on the condition that other unpatented machines be used by the lessee. This practice was condemned on the inference that competition is adversely affected when a supplier with substantial power in one market extends its power to a separate market through a tying arrangement. Once this theory was endorsed, tying cases became much easier to decide.

In its *International Salt* decision [*International Salt Co. v. United States*, 332 U.S. 392 (1947)], the Supreme Court ruled that a tying arrangement was a per se violation of the antitrust laws. International Salt owned patents on two salt-dispensing machines. The Lixator and the Saltomat were leased to industrial users on the condition that the lessees purchase all of the unpatented salt and salt tablets that would be used in the leased machines from International Salt. Thus, International Salt had put explicit tying provisions in their leases. The Court did not appear to examine the market for either the tying good or the tied good. It inferred the existence of market power from the existence of a patent on the tying machine. Then the Court intoned "it is unreasonable, *per se*, to foreclose competitors from any substantial market" (*International Salt*, p. 396) which introduced the quantitative substantiality test. In this instance, it was satisfied by sales of some $500,000 annually without regard to the size of the total market. Although the Court had no idea what percentage of the industrial salt market had been affected by International Salt's leases, it ruled that "the volume of business affected by these contracts cannot be said to be insignificant or insubstantial" (*International Salt*, p. 396). Thus, the rule enunciated by the Court in International Salt is that "foreclosure of a nontrivial volume of commerce provides sufficient basis for per se illegality."[5]

In 1953, the Supreme Court laid down the different standards of proof of illegality under the Sherman and Clayton Acts. In *Times-Picayune Publishing Co. v. United States* [345 U.S. 594 (1953)], the Court had to rule on the legality of a unit advertising plan. The *Times-Picayune* was the only morning newspaper in New Orleans. The Times-Picayune Publishing Co. also owned an afternoon newspaper, the *States*. The publisher required advertisers to purchase space in the morning and the evening newspaper to obtain either.

The Government challenged this unit advertising plan under Section 1 of the Sherman Act. The Court held that the Sherman Act condemns a tying arrangement whenever (a) "sufficient economic power" is shown in

---

[5] See Baker (1980, p. 1285.) This perceptive article offers a critique of the judicial analysis of each landmark tying case. On *International Salt* in particular, see Peterman (1979).

the tying good and (b) a "not insubstantial" amount of commerce in the tied good is affected (*Times-Picayune*, p. 608). In contrast, the Clayton Act is offended when either of these conditions is satisfied. Subsequent decisions have so attenuated the requirements for proving either condition that there is practically no distinction between the Sherman Act and the Clayton Act.[6]

The contemporary rule on tying was formulated in the *Northern Pacific* decision [*Northern Pacific Railway Co.* v. *United States*, 356 U.S. 1 (1958). For an interesting analysis of this decision, see Cummings and Ruhter (1979).] To encourage railroad expansion, Congress had given Northern Pacific some 40 million acres of land in a checkerboard pattern along its line. This allowed intermittent private development and subsequent sale or lease by the railroad. As Northern Pacific sold or leased its land, it included so-called preferential routing clauses in the agreements. These clauses provided that the lessee would ship all commodities produced or manufactured on the land over Northern Pacific's lines as long as Northern Pacific's rates were equal to those of competing carriers. The Court found these preferential routing clauses to be per se violations of the antitrust laws.

Justice Black explained the Court's hostility toward tying arrangements. First, competition on the merits in the tied-good market is attenuated when tying arrangements are imposed. Competitors are denied "free access to the market for the tied product, not because the party imposing the tying requirements has a better product or a lower price but because of his power or leverage in another market" [356 U.S. 1, 6 (1958)]. Second, "buyers are forced to forego their free choice between competing products" [356 U.S. 1, 6 (1958)]. Finally, tying arrangements confer no beneficial consequences on society to compensate for their deleterious effects. This point was made forcefully by Black's quoting one of the best-known judicial assessments of tying arrangements, Justice Frankfurter's opinion in an exclusive dealing case: "Tying arrangements serve hardly any purpose beyond the suppression of competition." [*Standard Oil Co. of California (Standard Stations)* v. *United States*, 337 U.S. 293, 305–306 (1949).] Since Justice Black could not find much to recommend tying arrangements, he wrote that tying arrangements "are unreasonable in and of themselves whenever a party has sufficient economic power with respect to the tying product to appreciably restrain free competition in the

[6] For example, in *Northern Pacific Railway Co.* v. *United States*, 456 U.S. 1 (1958), the Supreme Court found sufficient economic power in the unique and strategic location of the railroad's land (the tying good). The existence of a copyright on the tying good provided sufficient economic power in *United States* v. *Loew's, Inc.*, 371 U.S. 38 (1962). Moreover, in *Loew's*, a dollar volume of some $60,800 was found to be not insubstantial.

market for the tied product and a 'not insubstantial' amount of interstate commerce is affected'' [356 U.S. 1, 6 (1958)].

The Court's decision in *Loew's*[7] followed axiomatically from the *Northern Pacific* decision. This case involved the block booking of copyrighted movies for television exhibition. Since inferior films were accepted in order to get desirable pictures, market power was apparent. For example, in order to get a good film like "The Man Who Came to Dinner," one might have to purchase "Tugboat Annie Sails Again."

This case served to reduce the threshold standard for sufficient economic power in the tying product: "Even absent a showing of market dominance, the crucial economic power may be inferred from the tying product's desirability to consumers or from uniqueness in its attributes" [371 U.S. 45 (1962)]. Because the films were copyrighted, the Court inferred the existence of the requisite economic power. Consequently, the Court felt that determining the relevant market was not necessary.[8]

Finally, we reach the protracted litigation resulting from the dispute between Fortner Enterprises and U.S. Steel. (Fortner filed his case in 1962. Before getting a final Supreme Court decision in 1977, he made three trips to the District Court, three to the Court of Appeals, and one to the Supreme Court in 1969.) Fortner was a land developer in Kentucky who purchased some prefabricated homes from U.S. Steel. In addition, he obtained loans from the U.S. Steel Homes Credit Corp., a wholly owned subsidiary of U.S. Steel. This financing was alleged to be unique because it covered the cost of the homes plus the cost of the land. Consequently, Fortner obtained financing that covered more than 100% of the homes purchased. Given the Credit Corp.'s mission, the loans were conditioned on Fortner's purchasing the homes from U.S. Steel.

The *Fortner I* decision [*Fortner Enterprises, Inc.* v. *United States Steel Corp.*, 394 U.S. 495 (1969)] was ludicrous. At this point in the litigation, the question was whether U.S. Steel had the appreciable market power necessary to find the credit tie illegal. The Court held that appreciable economic power exists if "the seller has the power to raise prices or to impose other burdensome terms such as a tie-in, with respect to any appreciable number of buyers within the market" (*Fortner*, p. 504).[9] This logic would seem to make tying a per se offense without condition. This

---

[7] *United States* v. *Loew's, Inc.*, 371 U.S. 38 (1962). An interesting analysis of the *Loew's* decision is provided by Stigler (1963) who explains block booking as an effort to practice price discrimination.

[8] See Baker (1980) at 1241: "*Loew's* indicates the Court's permissive attitude toward the evidence necessary to satisfy the market power requirement."

[9] See Baker (1980) at 1242–1243 for a great dissection of the Court's standard for determining the existence of appreciable economic power.

decision essentially said that the ability of a seller to impose a tying arrangement on its customers constituted proof of the market power that makes tying illegal. But if this were the case, promotional sales would constitute illegal ties. As a reductio ad absurdum, for example, suppose a gasoline station offered a free car wash with the purchase of at least ten gallons of gasoline. This would be an illegal tying sale if a large number of customers took advantage of the offer.

When a much different Supreme Court had a second chance in *Fortner II* [*United States Steel Corp.* v. *Fortner Enterprises, Inc.* 429 U.S. 610 (1977)], the specific decision made some sense. The Court found that the Credit Corp's relationship with U.S. Steel did not confer any cost advantage. Since one U.S. Steel house was sold per lot financed, the tying took place in fixed proportions and, therefore, the large number of transactions did not signify anything. Although U.S. Steel charged $400 more per house than its competitors, the noncompetitive price simply meant that a lower price was paid for financing. Finally, if the evidence merely shows that credit terms are unique because the seller is willing to accept a lesser profit—or to incur greater risks—than its competitors, that kind of uniqueness will not give rise to any inference of economic power in the credit market (*Fortner II,* p. 622). The Court went on to point out that "the unusual credit bargain offered to Fortner proves nothing more than a willingness to provide cheap financing in order to sell expensive houses" (*Fortner II,* p. 622). Jones (1978) has interpreted *Fortner II* as requiring competitors to compete on the same terms. In other words, if other sellers of prefabricated homes feel foreclosed, they should offer the same sort of bargain on credit as that offered by U.S. Steel. Thus, the Court has made it a bit more difficult to sustain the burden of proof necessary for finding a combination sale to be a per se violation. Thus, our hypothetical gasoline station must have some unique advantage in offering car washes or other gasoline stations will be unable to show market power in the tying good.

Public Policy Assessment

We have seen that fixing maximum-resale prices, resale price maintenance, and tying arrangements have been accorded the ignominius status of per se illegality under the antitrust laws. But the advisability of per se treatment for these business practices can be questioned. As a matter of fact, the Supreme Court's own reasoning has provided some demanding standards for per se rules of illegality. For example, in *Northern Pacific,* Justice Black said that "there are certain agreements or practices which because of their pernicious effect on competition and lack of any redeeming virtue are conclusively presumed to be unreasonable and therefore illegal without elaborate inquiry as to the precise harm they have caused or the business excuse for their use" [356 U.S. 1, 5 (1958)]. In a more recent case, the Court expressed the view that "*per se* rules of illegality are appropriate only when they relate to conduct that is manifestly anticompetitive" [*Continental T.V., Inc.,* v. *GTE Sylvania,* 433 U.S. 36, 46 (1977)].

In a footnote, the *Sylvania* court went on to point out:

> *Per se* rules thus require the Court to make broad generalizations about the social utility of particular commercial practices. The probability that anticompetitive consequences will result from a practice and the severity of those consequences must be balanced against its procompetitive consequences. Cases that do not fit the generalization may arise, but a *per se* rule reflects the judgment that such cases are not sufficiently common or important to justify the time and expense necessary to identify them. Once established, *per se* rules tend to provide guidance to the business community and to minimize the burdens on litigants and the judicial system of the more complex rule of reason trials, . . . but those advantages are not sufficient in themselves to justify the creation of *per se* rules. If it were otherwise, all of antitrust law would be reduced to *per se* rules, thus introducing an unintended and undesirable rigidity in the law. (*Continental T.V.*, p. 46)

Given the Court's own standards for imposing the per se brand on a business practice, we have serious reservations about the per se label on the practices analyzed in this chapter. First, there is absolutely nothing that commends the per se illegality of fixing maximum-resale prices. An upstream firm with market power will attempt to impose maximum-resale prices whenever a successive monopoly situation is present. The purpose of the practice is purely selfish—to increase the upstream producer's profits. No one pretends that the producer is motivated by undiluted altruism. On the contrary, the producer is unabashedly motivated by greed. In this instance, however, the interests of the firm and the consumer coincide. Fixing maximum prices promotes consumer welfare. Consequently, a simple rule exists for handling instances of fixing maximum prices. Whenever the industry structure is one of successive monopoly, constraints on downstream pricing discretion should be permitted.

Resale price maintenance, that is, fixing minimum-resale prices, is not quite so easy to dismiss. As we have seen, there are a couple of anticompetitive possibilities associated with resale price maintenance. But there is also a procompetitive rationale for resale price maintenance. Consequently, it is not correct to label this business practice "manifestly anticompetitive." Judicial expediency, therefore, does not justify assigning a per se label to the resale price-maintenance agreement. At least as a first approximation, the firm should be granted a rebuttable presumption of legality if the resale price maintenance is followed by an expansion of output. This, of course, does not remove all ambiguity because demand may be declining or production costs may be increasing so that output would decline in any event. But it would be a step in the right direction.

Finally, we turn to tying arrangements. The appropriateness of per se treatment for tying has recently been questioned by several astute commentators. Milton Handler, for example, remarked that "There are second thoughts today on whether, as a matter of policy, we may not have gone too far in condemning tie-ins. Combination sales . . . may be an effective way of waging competition without any serious anticompetitive effects" (See Handler, 1977, p. 1019). Bork had a more forceful assessment: "Antitrust treats them [tying arrangements] as utterly pernicious, despite the increasingly obtrusive fact that it has found no adequate grounds for objecting to them at all" (see Bork, 1978b, p. 365). His dissatisfaction is consistent with that of Posner (1976) and (1977) and Jones (1978).

For many years, the use of tying arrangements has been a hazardous proposition. Most of the judicial hostility toward tying can be attributed to the leverage theory, which holds that firms with market power in one market can use tying to lever that power into the tied good market. Bowman (1957) has examined this theory carefully and found it wanting. As an alternative to the leverage theory, the so-called "Chicago" analysis of tying finds that the practice is apt to be used as a form of meter pricing for purposes of price discrimination. Since the overall social welfare effects of successful price discrimination are indeterminate on a priori grounds, this theory leads one logically to the advocacy of a rule of reason treatment of tying arrangements. In Chapter 3, we have proved an economic theory of tying that was earlier described by Burstein (1960) in which the practice is employed as a contractual alternative to vertical integration in situations where an intermediate-product monopolist sells its output to a competitively structured final-good industry that combines inputs in variable proportions. Once again, the welfare effects are found to be indeterminate, and the rule of reason approach is recommended. Like the earlier argument favoring a more lenient attitude toward tying, however, this recommendation has its logical foundation in the indeterminacy of the social welfare effects that result from the practice. Perhaps for this reason, the judiciary appears to have found it, too, less than totally convincing.

In Chapter 5, we have gone a step beyond demonstrating indeterminacy. Specifically, we have examined a situation in which the welfare effects of tying are both determinate and positive. This involved the reduction of uncertainty and the consequent expansion of output. If antitrust policy is really concerned with consumer welfare, then tying arrangements should be vigorously applauded in these circumstances.

9

Contractual Controls That Are Not Illegal Per Se

Introduction

In contrast to vertical price-fixing and tying arrangements, which are illegal per se, there are several vertical control mechanisms that are presumptively legal or at least subject to the rule of reason. We have classified output and sales revenue royalties along with lump-sum entry fees as presumptively legal. For the most part, an input monopolist (or franchisor) that extracts his payment in this way will not be subject to antitrust challenge. Things are not so sanguine, however, when it comes to territorial allocations, exclusive dealing, and requirements contracts. These practices have been challenged under the antitrust laws in the past. Currently, they are examined under the rule of reason, which provides the defendant with an opportunity to defend himself.

In this chapter, we shall examine these two groups of control devices. This examination will reveal the hazards facing an input supplier that is trying to extract monopoly rents through these vertical control mechanisms. As we shall discover, these hazards are closely related to the extent of market foreclosure involved.

Presumptively Legal Controls

In Chapter 5, we analyzed the economic equivalence between vertical-ownership integration and output royalties, sales revenue royalties, and lump-sum entry fees. We also discussed some of the practical problems of using these controls in real situations. Now, we turn to the legal ramifications of these control mechanisms.

Lump-Sum Entry Fees

These are probably used most frequently in franchise situations. The franchisor extracts part of his monopoly rent through an initial franchise fee. In principle, all of the monopoly rent could be obtained through this fee. But practical considerations create a divergence of expectations between the franchisor and the franchisee that make it impossible to rely solely upon the franchise fee.[1] Nonetheless, a large point in favor of lump-sum entry fees is their presumptive legality. As far as we can tell, collection of a lump-sum fee for the privilege of using a patent or trademark is a perfectly legal means of extracting monopoly rent.

Output Royalties and Sales Revenue Royalties

These methods of extracting monopoly rents are close cousins. An output royalty is a fixed charge per unit of final output. It behaves like a per-unit tax to extract profit from the downstream producer. The sales revenue royalty is a fixed percentage of the downstream firm's total sales revenue. Thus, it acts like an ad valorem tax on the downstream firm. Both of these royalties can be used to extract all of the monopoly rents from the downstream firms, at least in principle. In practice, there are several problems, which were discussed in Chapter 5.

As far as the antitrust laws are concerned, neither of these vertical control mechanisms poses a problem. But either one can pose some difficulties under the patent-misuse doctrine, a useful discussion of which is provided in Note (1978) along with the Supreme Court's latest pronouncement. This arises primarily in patent licensing situations, but it can arise in the sale of patented inventions as well. The current patent law confers upon inventors an exclusive right to the result of their inventive efforts for a period of 17 years. In particular, the present patent law provides in part that "every patent shall contain . . . a grant to the patentee, his heirs or assigns, for the term of seventeen years . . . of the right to exclude others from making, using, or selling the invention throughout the United States" 35 U.S.C. §154 (1970). The purpose of the patent grant is to promote technical progress by offering a limited monopoly to the successful inventor. If a patent holder abuses his patent privilege, however, the courts will not protect him against infringements. The patent-misuse doc-

---

[1] This argument is spelled out in Blair and Kaserman (1982c). Divergent expectations cause some loss in potential monopoly rents for the franchisor.

trine was created by the judiciary to deal with patent holders who tied unpatented commodities to the patented item. For example, the Supreme Court denied protection against infringement in *Morton Salt Co. v. G. S. Suppiger Co.*, 314 U.S. 488 (1942). In this case, Suppiger had a patent on a salt dispensing machine which was leased to canners on the condition that they buy Suppiger's unpatented salt tablets. When Morton produced a similar machine, Suppiger sued for patent infringement. Due to Suppiger's patent misuse, the Court dismissed the suit against Morton. The Court ruled that "it is the adverse effect upon the public interest of a successful infringement suit, in conjunction with the patentee's course of conduct (the tying arrangement), which disqualifies him to maintain the suit" [314 U.S. 488, 494 (1942)].

The patent-misuse doctrine has been applied in instances where a patent holder has permitted the use of his patent in exchange for a royalty levied on the downstream firm's total sales revenue. In most cases, the patent was licensed for use. [For examples, see *Automatic Radio Manufacturing Co. v. Hazeltine Research, Inc.* 339 U.S. 827 (1950), and *Zenith Radio Corp. v. Hazeltine Research, Inc.*, 395 U.S. 100 (1969).] Usually, if there is an objection to this form of compensation, it is because the royalty applies whether or not the final product incorporates the benefit of the patent. For example, in *Zenith*, 1969, Hazeltine's license permitted Zenith to use any of its domestic radio and television patents in exchange for a percentage of all Zenith sales of radios and televisions. The royalty had to be paid during the life of the contract even if no Hazeltine patents were used.

There is no reason why the patent-misuse doctrine could not be applied to situations in which the patented commodity was sold on the condition that a royalty on total sales revenue would be paid to the upstream firm. If the royalty applied only on the sale of final goods that actually used the patented commodity as an input, there would appear to be no problem. Only in instances where the contract was similar to Hazeltine's would the issue of patent misuse arise. Our discussion of output and sales revenue royalties in Chapter 5 implicitly assumed that the patented input was actually used in the production of the final good.

Exclusive Dealing and Requirements Contracts

An exclusive dealing contract commits a buyer to deal only with a specific seller. For the duration of the contract, one supplier has the exclusive right to sell to the buyer involved. Exclusive dealing normally refers to business practices that do not involve tying arrangements, which are illegal per se. A requirements contract is a close relative. It commits a buyer to take all it needs of a certain input from a specific seller. It could also commit the seller to supply all of a buyer's needs. These two types of contracts have drawn antitrust attention but have not been deemed illegal per se. It is clear, however, that either contract forecloses part of the market from competitors for the duration of the contract. This is the reason that these contracts have evoked antitrust concern.

Economic Motivation

Exclusive dealing and requirements contracts often promote efficiency and thereby enhance the profits of both buyer and seller. First, the seller and/or the buyer may experience cost reductions. The seller may need to call on fewer dealers, to keep fewer records, and perhaps experience fewer credit problems when opting for a smaller number of exclusive distributors as opposed to a larger number of nonexclusive distributors. Selling expenses may be greatly reduced or nearly eliminated by exclusive dealing. In addition, the supplier can plan his operations more effectively and experience efficiency gains in production as well as in sales efforts.

The buyer may also enjoy cost savings due to exclusive dealing. For one thing, the buyer can reduce the transaction costs of dealing with a multitude of competing suppliers. In addition, buyers may agree to exclusive dealing contracts in exchange for the guarantee of frequent small deliveries so that they can reduce their inventory costs.

Second, the supplier may get improved product promotion from those with exclusive contracts. There will be an added incentive to promote the seller's product vigorously if that is all the buyer has to sell to the final consumer. Thus, the supplier can be sure that each of the distributors will work very hard on the supplier's behalf. As a result, it will be worth providing special training to the distributor where the complex character of the product requires special skills.

Improved product promotion can be carried too far in a social sense. We can see this in Figure 9.1 where the original demand for and supply of the final good are denoted by D_1 and S_1. Competition among rival firms for market share may take the form of promotional efforts. At least to some extent, these efforts will be mutually offsetting. As a result, the shift in demand from D_1 to D_2 is apt to be smaller than the shift in supply from S_1 to S_2. We can see that the original quantity is worth P_2 per unit to the consumer, but its cost is now P_3. In a very real sense, the promotional efforts have been excessive and wasteful: the additional expenditures

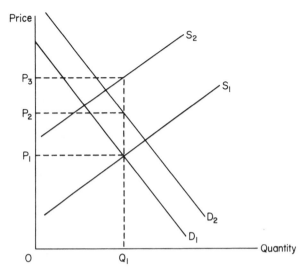

Figure 9.1 Socially excessive product promotion.

have exceeded the added value created by those expenditures. Areeda (1981, pp. 811–812) raised this issue in a somewhat oblique fashion. Stigler (1968) unintentionally made this point in another connection. Dixit and Norman (1978) provided a formal proof of this proposition.

Third, exclusive dealing or requirements contracts may be used to reduce risk. The supplier and/or the buyer may be risk averse. In that case, risk imposes costs, which the risk-averse firm will try to mitigate. If supply is uncertain, a risk-averse buyer will mitigate the effects of uncertainty by cutting back his planned output. This, in turn, reduces the derived demand for the supplier's output. Similarly, if demand for a supplier's output is random, his risk-averse response will be to plan on producing less. The buyers will be facing a reduced supply schedule. Thus, the presence of risk can make both buyers and sellers worse off. This, of course, means that each will be willing to pay a premium to reduce the risk that each is facing. (A more formal analysis is provided by Blair, 1974.)

Buyers may want the assurance of continuous supply to avoid the disruptions of periodic shortages. For example, an electric utility must have a stable and adequate supply of fuel in order to avoid having to interrupt service. If prices tend to be volatile, a buyer may want a specified price over some time interval for corporate planning purposes. In either event, the seller's quid pro quo may be an exclusive dealing or requirements contract. In order to avoid supply problems, buyers may be happy to agree to such an arrangement.

In summary, there are several reasons for buyers and sellers to embrace exclusive dealing contracts. The results of exclusive dealing when the motivation is to reduce costs or to reduce risk are apt to be beneficial for consumers. Even in the case of improved product promotion, the consumer may benefit. In practice, exclusive dealing has been used as a means of entry or as a way of propping up a weak product. In these two cases, the appropriate public policy response is fairly clear. When exclusive dealing facilitates the entry of a new brand which thereby intensifies competition in the final-good market or when exclusive dealing permits a relatively weak brand to survive and compete, there should be no antitrust prosecution. In both of these cases, the exclusive dealing contract intensifies the competitive process and, therefore, should be applauded.

Market Foreclosure

An exclusive dealing arrangement, by its very nature, must result in market foreclosure. This has an allegedly adverse impact upon the supplier's competitors. It might foreclose the market opportunities of the seller's competitors if the seller preempts all of the outlets. There is some concern that if one supplier captured all of the leading distributors, other producers would be severely handicapped in their efforts to reach consumers, even if their product were better or cheaper. Alternatively, if several leading producers entered into exclusive dealing arrangements with all of the distributors, new suppliers would find entry very difficult. Even existing suppliers would have a tough time expanding their share of the market. In other words, the requirements contract raises the barriers to entry for new firms and limits the expansion of existing firms.

The antitrust concept of monopoly is the power to exclude, that is, the power or ability to prevent competitors from gaining access to the market. Exclusive dealing or a requirements contract cannot create exclusionary power. It can only be a means of implementing such power as already exists. Thus, a requirements contract can only restrict competition in any meaningful sense when the power to exclude is already present. It should be clear that single exclusive dealing contracts are innocent. If such contracts are to pose a problem, it must be true that a seller who already has the ability to exclude insists upon a system of these contracts.

We must remember that exclusive dealing and requirements contracts are contractual forms of vertical integration. The objection to exclusive dealing and requirements contracts relies on the theory that competition may be injured through the market foreclosure of rivals when a firm vertically integrates by contract. If the foreclosure theory is fallacious as it pertains to ownership integration, then a fortiori it must be fallacious as it pertains to contractual integration. For example, suppose a seller attempts to monopolize its market by exclusive dealing. If there are no efficiencies, the seller's rivals will match whatever inducements are offered. Under these conditions, exclusive dealing cannot be exclusionary.

Suppose, however, that a superior product is offered at the same price or that a lower price is offered for the same quality product. Most people will agree that superior products or lower prices are to be applauded. Now, the seller must offer an inducement that exists for the life of the contract. Thus, in terms of excluding rivals, the contract offers the seller no advantage that it would not have had without the contract. If productive efficiency or superior products provide an advantage and an ability to exclude rivals, requirements contracts or exclusive dealing provide no added ability. Consequently, neither exclusive dealing nor requirements contracts will have much competitive effect.

Finally, we should recognize that these contracts are renewed periodically. The competition among sellers occurs at those times of contract renewal. Thus, competition does not disappear; it just takes place intermittently rather than continuously. It is not obvious why this poses any serious problems. This is especially true when a seller's exclusive dealing contracts have a wide range of renewal dates.

Legal Treatment

Neither exclusive dealing nor requirements contracts are illegal per se. As a result, they are subject to the rule of reason even though they do not usually get a full-blown rule-of-reason treatment. The courts have recognized explicitly that these contracts can have beneficial consequences. Accordingly, the courts have sought a means of easily distinguishing those instances in which the deleterious consequences predominate and those in which the beneficial consequences predominate. This search can be seen in the review of the Supreme Court cases.

One of the earliest cases is *Standard Fashion Co.* v. *Magrane-Houston Co.*, 258 U.S. 346 (1922). In this case, Standard Fashion made and distributed patterns for women's and children's clothes. Magrane-Houston was a retail dry goods store that sold these patterns among a multitude of other things. The two parties entered into an exclusive dealing contract that required Magrane to maintain a substantial inventory, to remain in the same location, and to sell only Standard's patterns at full list price. The price of the patterns to Magrane was 50% of the retail list price. Moreover, discarded patterns could be returned for 90% credit on new patterns. Finally, the term of the contract was two years. Before the contract expired, Magrane discontinued selling Standard's patterns and began selling McCall's patterns. Standard sued for enforcement of the contract.

Since the contract was quite explicit, the Supreme Court held that there was no question about whether exclusive dealing existed. The real question was whether this particular contract should be struck down because the covenant not to sell the patterns of others "may be to substantially lessen competition or tend to create a monopoly." The Court reasoned that "the Clayton Act sought to reach the agreements embraced within its sphere in their incipiency, and . . . to determine their legality by specific tests of its own." The Court's test turned out to be structural.

There were 52,000 pattern agencies in the entire country. A holding company controlled Standard and two other pattern companies, and through contractual arrangements, controlled about 40% of the pattern agencies. These facts led the Court to two inferences. First, in small towns, the pattern business is apt to be monopolized by a single pattern manufacturer. Second, in large towns and cities, exclusive contracts with the most attractive retailers will lead to monopoly. As a result, these exclusive dealing contracts were found to violate Section 3 of the Clayton Act.

The Court moved away from its structural test in its "Standard Stations" decision and toward the vertical market foreclosure test [*Standard Oil Co. of California* v. *United States,* 337 U.S. 293 (1949)]. In 1946, Standard Oil sold 23% of the total taxable gallonage of gasoline sold in the so-called Western Area, which consisted of Arizona, California, Idaho, Nevada, Oregon, Utah, and Washington. At the retail level, Standard Oil sold 13.5% of the total while its six leading competitors sold some 42.5%. All of these companies used similar exclusive dealing arrangements. Most of these contracts were for one year, but both parties envisioned a continuing relationship. Standard Oil had 5937 independent stations under exclusive supply contracts. This amounted to about 16% of all retail gasoline outlets in the area.

Between 1936 and 1946, Standard Oil's sales through independent dealers remained a relatively constant proportion of total sales. Since Standard Oil's contracts clearly involved exclusive dealing, the question was whether the contract had the effect of substantially lessening competition. The district court inferred a substantial effect. It felt that the requirement of showing an *actual* or *potential* lessening of competition or a tendency to establish monopoly was adequately met by proof that the contract covered a substantial number of outlets and a substantial amount of products, whether considered comparatively or not. Given such quantitative substantiality, the District Court reasoned that the substantial lessening of competition is an automatic result, for the very existence of such contracts denies dealers the opportunity to deal in the products of competing suppliers and excludes suppliers from access to the outlets controlled by those dealers.

Given the district court's ruling, the issue before the Supreme Court was obvious: was the burden of showing a substantial lessening of competition met by proof that a substantial portion of commerce had been affected? Alternatively, must it be shown that competitive activity has actually diminished or probably will diminish? This case was new because Standard Oil could not be considered to occupy a dominant position. But *International Salt*[2] rejected the necessity of demonstrating economic consequences once it had been established that the volume of business affected was neither insignificant nor insubstantial and that the effect was to foreclose competitors from a substantial market.

The Supreme Court explicitly recognized that requirements contracts may well be of economic advantage to buyers as well as to sellers and thereby may help the consumer. For the *buyer,* requirements contracts may assure supply, guarantee price, enable long-term planning on the basis of known costs, and reduce inventory problems. For the *seller,* they may substantially reduce selling expenses, protect against price fluctuations, and offer the possibility of a predictable market, thereby facilitating entry. In fact, the Court proposed various tests of economic usefulness: (1) evidence that competition flourished despite the use of the contracts, (2) the length of the contract relative to the reasonable requirements of the field of commerce in which they were used, (3) the status of the defendant as a struggling newcomer or as an established competitor, and perhaps most important, (4) the defendant's degree of market control.

Apparently Standard Oil's contracts failed these tests. The Court emphasized that it is the theory of the antitrust laws that the long-run advantage of the community depends upon the removal of restraints upon competition. The Court noted that there were alternative "ways of restricting competition" and remarked that as long as these remained available, "there can be no conclusive proof that the use of requirements contracts has actually reduced competition below the level which it would otherwise have reached or maintained." But Section 3 of the Clayton Act does not require *conclusive* proof of anticompetitive effect. "To interpret that section as requiring proof that competition has *actually* diminished would make its very explicitness a means of conferring immunity upon the practices which it singles out. Congress has authoritatively determined that those practices are detrimental where their effect *may be* to lessen compe-

[2] *International Salt Co.* v. *United States,* 332 U.S. 392 (1947) was a tying case brought under Section 3 of the Clayton Act. In that case, the Court ruled that market foreclosure of $500,000 was large enough to satisfy Section 3 without investigating the size of the market.

tition." The requisite probability was easily satisfied: "We conclude, therefore, that the qualifying clause of §3 is satisfied by proof that competition has been foreclosed in a substantial share of the line of commerce affected." The problem with the majority's approach is revealed in Jackson's dissent: "The number of dealers and the volume of sales covered by the arrangement of course was sufficient to be substantial. That is to say, this arrangement operated on enough commerce to violate the Act, provided its effects were substantially to lessen competition or create a monopoly. But proof of their quantity does not prove that they had this forbidden quality."

The Supreme Court's fairly stringent attitude continued with its ruling in *Motion Picture Advertising* [*Federal Trade Commission* v. *Motion Picture Advertising Service Co.*, 344 U.S. 392 (1953)]. The Motion Picture Advertising Service Co. (MPAS) made advertising films for exhibition in movie theaters. MPAS paid the theaters for showing their advertising films on the condition that the theater refuse to show anyone else's films. These exclusive arrangements were prevalent in the industry: MPAS and three of its competitors had tied up about 75 percent of the theaters that showed such films.

The Federal Trade Commission (FTC) challenged this arrangement as an unfair method of competition in violation of Section 5 of the Federal Trade Commission Act. As a result, the FTC ordered MPAS and the others to limit their contracts to 1 year or less on the ground that longer terms unreasonably restrain competition and tend to create monopoly.

The Supreme Court found that the FTC ruling was appropriate in this case: "The exclusive dealing arrangement which has sewed up a market so tightly for the benefit of a few falls within the prohibitions of the Sherman Act and is therefore an 'unfair method of competition' within the meaning of §5(a) of the FTC Act."

In its *Tampa Electric* decision, [*Tampa Electric Co.* v. *Nashville Coal Co.*, 365 U.S. 320 (1961)], the Supreme Court appeared to relax its views and justified its ruling through a clever market definition. There are reasons, however, for not jumping to the conclusion that this decision signals a general relaxation of hostility toward exclusive dealing and requirements contracts. In 1955, Tampa Electric decided to expand its facilities by constructing an additional generating plant with six generating units. Tampa Electric decided to use coal as fuel in the first two units at the Gannon Station. Concerned about having a ready supply of coal for the life of the generating units, Tampa Electric entered into a requirements contract. The supply contract called for Tampa Electric to obtain its "total requirements of fuel . . . not less than 225,000 tons of coal per unit year" for a period of 20 years from Nashville Coal. As Tampa Electric constructed other units at Gannon Station, they would be added to the contract unless they were not designed to burn coal. The minimum price for the coal was $6.40 per ton delivered, subject to an escalation clause based on labor cost and other factors.

When Nashville Coal refused to honor its contract, Tampa Electric had to find another source. The terms were not as good and the price was some $8.80 per ton. Tampa Electric's estimated requirements for the first unit were 350,000 tons in 1958, 700,000 tons in 1959 and 1960, 1,000,000 tons in 1961, and steady increases until an annual average of 2,250,000 tons was reached.

The Supreme Court recognized several structural facts in this industry. First, total coal consumption in peninsular Florida other than Tampa Electric was 700,000 tons per year. Second, there were 700 coal suppliers in the Nashville Coal-producing region. Finally, Tampa Electric's demand for 2,250,000 tons amounted to some 1% of the total production in the Nashville Coal-producing region.

With respect to the business practice of exclusive dealing, the Court remarked that "(a)n exclusive dealing arrangement . . . does not violate the section unless the court believes it probable that performance of the contract will foreclose competition in a substantial share of the line of commerce affected." In order to determine the effect, the Court pointed out that the line of commerce, the geographic market, and the amount of commerce foreclosed would have to be determined. The threatened foreclosure of competition must be judged in relation to the market affected. In other words, the competition foreclosed by the contract must be found to constitute a substantial share of the relevant market. Merely showing that the contract itself involves a substantial number of dollars is ordinarily of little consequence. This, of course, is in marked contrast to its prior and subsequent rulings in tying cases where a large number of dollars is deemed sufficient to satisfy the notion of quantitative substantiality.

For purposes of analysis, the Court *assumed* that the contract involved exclusive dealing and that the line of commerce was bituminous coal. The geographic market was the area that was served by the Appalachian coal area producers. The volume of coal affected was only .77% of total production and, therefore, in the relevant market was "quite insubstantial." The Court remarked that

> although protracted requirements contracts may be suspect, they have not been declared illegal per se. Even though a single contract between single traders may fall within the initial broad proscription of the section, it must also suffer the qualifying

disability, tendency to work a substantial—not remote—lessening of competition in the relevant competitive market. There is here neither a seller with a dominant position in the market as in *Standard Fashion;* nor myriad outlets with substantial sales volume and an industry-wide practice as in *Standard Stations.*

Moreover, the 20-year period appeared necessary to assure a steady and ample supply of fuel.

Assessment of Legal Doctrine

It is clear that prior to *Tampa Electric* the law on exclusive dealing and requirements contracts expressed a judicial concern for vertical foreclosure of markets and restrictions on freedom of choice. This concern is misplaced. First, we have seen that the foreclosure theory cannot support a prohibition of exclusive dealing because the theory is fallacious. Second, freedom of choice is not constrained by these contracts in any meaningful sense. Both parties reach agreement in advance of signing the contract. Prior to signing, the parties examine the available options and freely choose what each perceives to be in its self-interest. In spite of these observations, there is a lingering concern that a large supplier can coerce its distributors to accept an exclusive dealing arrangement. This residual concern is also unfounded.

If a supplier has a particularly desirable product that most distributors want to sell, he can extract the value of that product in the form of higher prices. Alternatively, he can impose exclusive dealing on his distributors. But he cannot do both. Consequently, it is apparent that the supplier cannot get something for nothing. If he wants exclusive dealing, he must purchase it from the distributors in the form of lower prices or the provision of greater efficiencies. The supplier who purchases exclusivity must do so because its costs are reduced as a result. This follows because exclusive dealing cannot be used as a vehicle for monopolization.

Tampa Electric contained a structural analysis and signalled a departure from the earlier reasoning. But this case had some special circumstances. In particular, Nashville Coal was trying to renege on a contract that it had entered into freely, because it regretted the terms. Thus, one of the parties was trying to escape the consequences of breaching a contract through the antitrust laws. Bork has pointed out that a similar amount of foreclosure subsequently was found to violate the law. [See Bork (1978, 302–303) for a discussion of the Court's decision in *Federal Trade Commission* v. *Brown Shoe Company,* 384 U.S. 316 (1966).] Thus, *Tampa Electric* cannot be viewed as a fundamental change in direction. Antitrust policy tends to be generally hostile to exclusive dealing.

Territorial Restraints

There is a host of nonprice vertical restraints that fall under the heading of territorial restrictions.[3] For the most part, these restraints are alternatives—albeit somewhat imperfect ones—to resale price maintenance.[4] Not surprisingly, the things that motivate resale price maintenance also motivate vertical territorial restraints. We shall consider three purposes served by territorial restraints. First, an input monopolist may have to provide protection for the downstream firm. When the input monopolist has charged a lump-sum entry fee to the downstream firm, it has collected the monopoly rent in advance. (See the discussion of lump-sum entry fees in Chapter 4.) The downstream firm must be able to charge the monopoly price for the final good in order to recover the entry fee and earn a normal return on its investment. In order to collect the capitalized value of the monopoly rents in advance, the upstream monopolist must provide some assurances that the downstream firm will not be confronted by competitors. Without such assurances, the downstream firm will not pay the fully capitalized value of the future monopoly rents.[5] The upstream firm cannot promise that there will be no interbrand competitors, but it can promise to insulate the downstream firm from intraband competitors. In order to do this, it will have to impose territorial restraints on all of the firms that have obtained permission to use its input. Otherwise, it will be profitable for one downstream firm to invade the market of another. In the absence of retaliation, the invader will be able to earn more than a competitive return. Thus, each downstream firm must be confined to its assigned territory. If this is accomplished, the price and output of the final good will be precisely the same as it would have been if the upstream firm had vertically integrated forward.

Second, the upstream firm may protect the downstream firms from free riders so that the local sales effort will be optimized.[6] These efforts may include local advertising as well as the provision of commodity-specific presale services. Without an exclusive territory, the local dealer may not be able to recover its investment in sales promotion because other dealers

---

[3] A companion group of restraints is called customer restrictions. Analytically, these amount to the same thing, so we shall deal with territorial restraints in the text.

[4] One of Bork's major contributions was to develop thoroughly the economic equivalence of market division and vertical price-fixing. See Bork (1966).

[5] Bierman and Tollison (1970) have discussed the difficulties of fully capitalizing monopoly rent. Blair and Kaserman (1982b) have shown that a franchisor's profits are smaller when it cannot fully assuage the franchisee's concern about the certainty of future demand.

[6] This is discussed in somewhat more detail in Chapter 8 along with resale price maintenance.

are making the sales. These other dealers are able to reduce their prices because they do not incur large sales promotion costs. Instead, they let other firms incur those costs and they take a free ride. The result of this is that dealers will provide less commodity-specific presale services and sales effort than the manufacturer would find optimal. By providing exclusive territories, each dealer will be secure in the knowledge that it will enjoy all of the benefits of expenditures on presale promotion and services.

Comanor (1968) was skeptical about the social benefits of local sales efforts. He recognized that dealer provision of advertising, demonstrations, and credit may be necessary and useful, but "the need for vertical restraints to ensure their provision indicates only an insufficient demand for them in an unrestricted market." He went on to object that the system encourages the joint supply of products plus services at a price in excess of the sum of the prices of the separate components. Somehow, joint supply is supposed to lead to an increase in market power, but this synergy is not articulated clearly. In some sense, Comanor seemed to be concerned with a problem that is somewhat different from ours. We are dealing with the efficient exploitation of market power whereas Comanor seemed to be dealing with the creation of market power.

Comanor dismissed the free-rider problem. After examining the argument, he concluded that "to the extent that services are demanded by consumers, a market will develop to supply them and a separate price will be charged." Moreover, if "manufacturers have a legitimate interest in having them provided, they should be forced to bear the cost." Under these circumstances, no vertical restraints are required. Whether Comanor's suggestion is sensible appears to be an empirical question.

Finally, the supplier may impose vertical territorial restraints to minimize the cost of postsale service and reduce the probability of customer dissatisfaction. This is particularly important for complex commodities: automobiles, computers, sophisticated office equipment, and so on. The main problem posed by the manufacturer's desire for adequate postsale service is, again, the free rider. Distant dealers may cut the price and invade the natural market of a particular dealer. They can afford this price cutting because the buyer will not require postsale services from them. Instead, the buyer will expect to receive postsale services from the local dealer. Thus, the distant dealer will want to enjoy a free ride on the local dealer. The local dealer, however, will not be anxious to provide unremunerative postsale services without having collected the profit on the original sale. Consequently, consumers will not receive adequate postsale service, to their detriment and to that of the manufacturer as well.

An exclusive territorial assignment helps to correct this situation by assuring the local dealer that he will be able to charge for the postsale services in the original sales price. He will have an incentive to keep the customer satisfied so that he will not switch brands. There are alternatives to closed territorial assignments: (1) selling postsale service separately and (2) requiring that the nearest dealer provide the service. Both of these are subject to abuse. See Bork (1966, pp. 447–448) for details.

In summary, we can see that the main motivations for vertical territorial allocation have nothing to do with the creation of monopoly power. When some market power exists, vertical territorial allocation permits increased effectiveness in the exploitation of that power. It cannot, however, be used to create additional power. For the most part, territorial allocations serve the same purposes as resale price maintenance. Neither should suffer particularly hostile antitrust treatment.

Judicial Treatment of Territorial Restraints

At the outset, we must distinguish carefully between horizontal and vertical territorial restraints. When market division is imposed horizontally, it will be seen as an alternative to horizontal price-fixing, which is illegal per se. The obvious examples have been well settled for quite some time. For example, in *Addyston Pipe & Steel*, [*United States* v. *Addyston Pipe & Steel Co.* 85 Fed. 271 (6th Circuit), affirmed 175 U.S. 211 (1899)], specific cities were assigned to certain manufacturers of cast iron water pipe. This horizontal market division was condemned as a violation of Section 1 of the Sherman Act.

Some less obvious cases are more troublesome. In *Topco*,[7] for example, the Court was faced with a horizontal territorial division among independent grocery stores. The purpose of the organization was to act as a cooperative buying association. This would provide the buying efficiencies of the large chains without actual ownership integration of the separate stores. Consequently, the Topco arrangement provided a means of entering the industry on a small but efficient scale. The antitrust conflict was between the enhanced interbrand competition that Topco facilitated and the reduced intrabrand competition that the restrictions caused. The Court ruled against Topco: "We think that it is clear that the restraint in this case is a horizontal one, and therefore, a per se violation of §1." For good or ill, the antitrust law has been applied consistently in horizontal cases.

For vertical territorial restraints, the law has not been applied so con-

---

[7] *United States* v. *Topco Associates*, 405 U.S. 596 (1972). There are other examples of this hostility toward horizontal market division that may not, on balance, be anticompetitive. For example, see *United States* v. *Sealy*, 388 U.S. 350 (1967).

sistently. Starting in 1948, the Antitrust Division asserted that vertical territorial restraints were per se illegal. Rather than risk going to trial, numerous firms negotiated consent decrees with the Antitrust Division that precluded the use of territorial restraints. But there was little explicit case law until *White Motor* [*United States* v. *White Motor Company,* 372 U.S. 253 (1963)]. White's agreement with its dealers provided that they would sell trucks only to customers located within their assigned territories. In exchange, White gave exclusive distributorships to its dealers. White argued that interbrand competition would be fostered by the dealer system, which required the restrictions. The government argued that vertical market division was analogous to horizontal market division and, therefore, was per se illegal. The lower court granted summary judgment in favor of the Department of Justice, a decision the Supreme Court felt was improper. Specifically, the Court held that "the applicable rule of law should be designed *after* a trial." Moreover, "[w]e know too little of the actual impact . . . of (a territorial) restriction . . . to reach a conclusion on the bare bones of the documentary evidence before us."

Per se violations involve practices which are presumed to be unreasonable and therefore illegal without elaborate inquiry as to the precise harm caused by their pernicious effect on competition and lack of any redeeming virtue. The Court would not categorize vertical restrictions as inherently pernicious, due to its lack of experience with them.

In Brennan's concurring opinion, he pointed out that restrictions on intrabrand competition that may promote interbrand competition should be weighed in court for their net effect.

Shortly after confessing an inability to fashion a rule of law on vertical territorial restraints, the Court was confronted by the *Schwinn* case [*United States* v. *Arnold, Schwinn & Co.,* 388 U.S. 365 (1967)]. In 1951, Schwinn had a 22.5% share of the U.S. market for bicycles. Faced with foreign and domestic competition, Schwinn's share had eroded to less than 13% by 1961. Although Schwinn was declining relatively during the 1951–1961 decade, it was growing absolutely. The company employed a complex distribution scheme. First, it sold bicycles to 22 wholesale distributors who had been granted territorial exclusives. There were restrictions on the resale of those bicycles: each distributor was limited to its own territory and could only sell to Schwinn franchisees. Second, Schwinn sold to retailers on a consignment basis with the same distributors and the same resale restraints. Finally, and quantitatively most important, Schwinn used its Schwinn Plan where the bicycles were sold directly to its retailers with commissions going to the distributors. All of the Schwinn franchisees were free to sell other brands but were expected to give Schwinn bicycles equal prominence. The franchisees were permitted to sell only to ultimate customers—not to discount houses. These customer and territorial restrictions were enforced by Schwinn through threats of termination.

The Court claimed that it was engaging in a rule-of-reason analysis. Although it recognized Schwinn's contention that its distribution system enhanced interbrand competition, the Court said that "the antitrust outcome does not turn merely on the presence of sound business reason or motive. Our inquiry is whether . . . the effect upon competition in the market place is substantially adverse." Irrespective of this assertion, the Court's decision did *not* depend upon an analysis of competitive impact. The Court drew an artificial distinction between a case in which the manufacturer parts with title and a case in which it retains ownership. Restrictions upon resale once title had passed were deemed per se violations of Section 1 of the Sherman Act. This came as a great shock after the Court's candid admission in its *White Motor* decision.

For 10 years, academic commentary on *Schwinn* was severely critical. In addition, in the post-*Schwinn* era, the lower courts were trying to sidestep the harsh per se rule of *Schwinn* by distinguishing the facts in the cases under consideration. One of these cases was *Sylvania,* which involved a location clause [*Continental TV* v. *GTE Sylvania,* 433 U.S. 36 (1977)].

GTE Sylvania was a rather small producer of television sets which were sold through a network of franchised retailers. These retailers were permitted to sell only from locations specified by Sylvania. Its careful selection of highly competent retailers resulted in a large increase in market share from 1–2% to about 5% for Sylvania. Continental was one of Sylvania's franchisees. Following a dispute over the location clause, Continental filed suit against Sylvania alleging that the location clause was a per se violation of Section 1 under the *Schwinn* rule.

The Ninth Circuit Court of Appeals drew a distinction between the *Sylvania* facts and the *Schwinn* facts. It decided that Sylvania was entitled to a rule of reason treatment and ruled in favor of Sylvania.

The Supreme Court recognized that the proverbial distinction without a difference had been drawn by the Ninth Circuit.[8] There was no substantive difference between Schwinn's restraints and those of Sylvania. Nevertheless, the *Schwinn* rule would not be applied because *Schwinn* was being overruled. The Court reviewed the demanding standards for per se illegality and decided that vertical restraints simply did not measure up. Vertical restraints have procompetitive effects on interbrand competition

---

[8] "Unlike the Court of Appeals, however, we are unable to find a principled basis for distinguishing *Schwinn* from the case now before us" (*Sylvania,* p. 42).

at the same time that they inhibit intrabrand competition. These conflicting effects cannot sensibly be resolved with a per se rule. The current rule is that *nonprice* vertical restraints will be judged under the rule of reason.

In summary, the law on territorial restraints has bounced around a little bit from rule-of-reason to per se and back again. One cannot be sure that the rule will not change again. Now, at least, the antitrust treatment makes some sense. The procompetitive and anticompetitive consequences of a vertical restraint will be weighed and judged according to the rule of reason. This, of course, does not mean that all restraints will pass muster, but at least they will be given a chance.

Concluding Remarks

In this chapter, we have pointed out the apparent legality of some vertical contractual controls: lump-sum fees, output royalties, and sales revenue royalties. Aside from the possible patent-misuse issue, these controls seem to be perfectly legal. In contrast, exclusive dealing, requirements contracts, and territorial restrictions may provide some problems. These controls are subject to rule-of-reason analysis on a case-by-case basis. There have been instances where these vertical controls have been found to violate the antitrust laws. Consequently, a firm must be cautious in using such controls in lieu of ownership integration. We must emphasize, however, that the economic foundation for antitrust hostility resides in the market foreclosure doctrine. Since this is fallacious, the mistreatment of these vertical controls is suspect.

Part III

Public Policy Analysis

10

Conclusion

We have seen that most vertical controls restrain the freedom of ostensibly independent firms, which would seem to make them vulnerable to antitrust attack as restraints of trade. But all contracts restrain freedom to some extent. What is important is the impact on competition and consumer welfare. For the most part, the antitrust law has condemned all concerted horizontal arrangements that have the purpose or effect of reducing competition. In contrast, most arrangements that are adopted for other purposes will not be condemned merely because of some incidental or inconsequential restraint on competition.

A similar attitude should guide public policy toward vertical arrangements. Unfortunately, this is not the case. Present policy concerning vertical restraints is both overly harsh and inconsistent. Our suggestions for policy changes are based on the fundamental purposes served by the alternative legal rules. The following discussion describes these purposes and the situations in which a given rule is preferred.

Alternative Legal Rules

Per Se Rules

If a business practice is per se illegal, the plaintiff need only prove that the practice occurred. Neither its actual motivation nor effect are relevant to the determination of antitrust liability. Recently, the Supreme Court reviewed the criteria that determine per se rules. [See *Continental TV Inc.* v. *GTE Sylvania,* 433 U.S. 36 (1977).] First, the Court expressed the view that "Per se rules of illegality are appropriate only when they relate to conduct that is manifestly anticompetitive" (*Continental* TV, p. 46). An earlier decision in a tying case provided some demanding standards for per se rules: "there are certain agreements or practices which because of their pernicious effect on competition and lack of any redeeming virtue are conclusively presumed to be unreasonable and therefore illegal without elaborate inquiry as to the precise harm they have caused or the business excuse for their use" [*Northern Pacific Railway Co.* v. *United States* 356 U.S. 1, 5 (1958)]. Although this decision held that tying arrangements fit this description, we have seen that it is incorrect to say that tying has no redeeming virtue.

In a footnote, the *Sylvania* court went on to point out:

> Per se rules thus require the Court to make broad generalizations about the social utility of particular commercial practices. The probability that anticompetitive consequences will result from a practice and the severity of those consequences must be balanced against its procompetitive consequences. Cases that do not fit the generalization may arise, but a *per se* rule reflects the judgment that such cases are not sufficiently common or important to justify the time and expense necessary to identify them. Once established, *per se* rules tend to provide guidance to the business community and to minimize the burdens on litigants and the judicial system of the more complex rule of reason trials, . . . but those advantages are not sufficient in themselves to justify the creation of *per se* rules. If it were otherwise, all of antitrust law would be reduced to *per se* rules, thus introducing an unintended and undesirable rigidity in the law. (*Sylvania*, footnote 1 *supra*, p. 46)

The Rule of Reason

According to *Sylvania,* the validity of vertical restraints will be judged by the rule of reason. A measure of skepticism may not be unwarranted, however, given the Supreme Court's historical inconsistencies.[1] Nonetheless, it is of some interest to determine what is meant by the rule of reason and how it should be implemented in vertical restriction cases. Prior to Chief Justice White's decision in *Standard Oil of N.J.* v. *United States* [221 U.S. 1 (1911)], the rule of reason was an underdeveloped manifestation of the doctrine on ancillary restraints.[2] The *Standard Oil* decision, however, provided a fairly clear interpretation of the Sherman Act. In particular, the statute was not designed to restrain the right to make and enforce contracts, but to protect competition from being restrained. As a result, Section 1's prohibitions were drawn to provide "an all-embracing enumeration to make sure that no form of contract or com-

---

[1] Bork (1977, p. 172) was led to remark: "It is entirely possible that *Sylvania's* promising start will be ended in a few years by another Schwinn. This law has bumped in the dark from one inconsistency to another since *Dr. Miles* in 1911, and so far it has always proved premature to announce the dawn."

[2] The classic statement was provided by Judge Taft in *United States* v. *Addyston Pipe and Steel Co.,* 85 Fed. 271 (6th Cir., 1898).

bination by which an undue restraint of interstate or foreign commerce was brought about could save such restraint from condemnation.'' Since the Sherman Act contains no explicit standard by which to determine the existence of a violation, Chief Justice White inferred "that the standard of reason . . . was intended to be the measure used for the purpose of determining whether, in a given case, a particular act had or had not brought about the wrong against which the statute provided.'' Of course, the standard of reason was necessary to distinguish *undue* restraints from inoffensive restraints. The sole factual issue before a court is whether the practice in question restricts competition to a degree that could be deemed undue. Chief Justice White did not suggest that a practice that significantly restricts competition can be saved by a finding that the public is better off in some other way. This makes good sense since the maintenance of competition is the overriding value that gives meaning to Section 1. Accordingly, it leaves little room for judicial discretion in weighing the values of competition against alternative social values.

Due at least in part to White's prolixity, the preferred quotation of the rule of reason standard is that of Brandeis. In *Chicago Board of Trade* v. *United States* [246 U.S. 231 (1918)], Brandeis improved upon the *Standard Oil* statement by highlighting the kinds of facts that may assist the courts in determining whether a particular restraint unduly restricts competition. After noting that all trade agreements necessarily restrain, Brandeis pointed out that:

> The true test of legality is whether the restraint imposed is such as merely regulates and perhaps thereby promotes competition or whether it is such as may suppress or even destroy competition. To determine that question the court must ordinarily consider the facts peculiar to the business to which the restraint is applied; its condition before and after the restraint was imposed; the nature of the restraint and its effect, actual or probable. The history of the restraint, the evil believed to exist, the reason for adopting the particular remedy, the purpose or end sought to be attained, are all relevant facts. This is not because a good intention will save an otherwise objectionable regulation or the reverse; but because knowledge of intent may help the court to interpret facts and to predict consequences [246 U.S. 231, 238 (1918)].

The court in *Sylvania* followed tradition by similarly quoting Brandeis with apparent approval, 433 U.S. 36, 45, n. 15 (1977). In spite of its popularity, one must agree with Posner that "[t]his is not a helpful formulation.'' (Posner, 1977, p. 15). The judiciary has simply failed to provide a truly manageable rule-of-reason standard that would permit a third party to predict court decisions.

A giant step forward would be taken by adopting Bork's (1965) long-held contention that the rule of reason should condemn those practices that are economically inefficient and condone those that are economically efficient. This would render the impact upon competition the only variable to be weighed in applying the rule of reason. As a consequence, every contract, combination or conspiracy that in purpose or probable effect will *significantly* restrain competition would be condemned by the rule of reason.[3] For some purposes, such a conception of the rule of reason is highly appropriate. But, as we have seen, many vertical restraints may have procompetitive as well as anticompetitive consequences. For these restraints, the rule of reason contemplated by Bork's suggestion would require the court to determine the net effect of the restraint. The requisite quantification would make many trials extremely complex. As a result, one should attempt to rely upon a priori determinations wherever possible, a recommendation reflecting the essence of Posner's (1977, pp. 16–20) proposal.

Appropriate Antitrust Policy

If we accept consumer welfare as the appropriate standard for analyzing the social desirability of a given business practice, then the antitrust treatment of vertical integration and its various contractual equivalents should be founded upon the consumer welfare effects implied by the existing models that explain such integration and control. Ignoring models in which the incentive to integrate stems from direct governmental action (e.g., the incentive provided by rate-of-return regulation) and grouping the various theories involving uncertainty into one broad category, Table 10.1 summarizes the welfare effects indicated by the theories we have surveyed.

The first thing that we notice in this table is that, of the theories represented, in only one case are the consumer welfare effects of vertical integration negative on a priori grounds. Moreover, as we saw in Chapter 3, the empirical importance, if not validity, of this particular theory is still the subject of some debate. Consequently, per se rules providing blanket prohibitions against vertical integration or its contractual equivalents are clearly inappropriate. Next, we find that in four cases the welfare effects of vertical integration are indeterminate without more specific information relating to the production and demand conditions present. Therefore, in these cases, a rule-of-reason approach would appear to be warranted. And finally, we see that in four more cases vertical integration and control generate unambiguous welfare gains. Here, a policy of per se legality is clearly appropriate.

[3] Sullivan (1977, pp. 186–189) feels that this is already the dominant modern conception of the rule of reason. Posner (1977, pp. 13–16), however, does not appear to be convinced.

TABLE 10.1

Economic Theories of Vertical Integration and
Implied Welfare Effects

Theory	Welfare Effects
Transaction costs	+
Successive monopoly	+
Product-specific services	?
Entry barriers	−
Variable proportions	?
Uncertainty	?
Monopsony	+
Price discrimination	?
Disequilibrium	+

Thus, if the costs involved in carrying out a full-blown rule-of-reason analysis are not too severe, the results of our survey suggest a policy approach that attempts to categorize observed vertical strategies according to the above set of competing theories. Given such a categorization, then, two major groups emerge, one of which should be subject to a rule of reason approach. If, on the other hand, the costs of carrying out the analysis required for a careful rule of reason treatment are prohibitively high, then a "second-best" policy would be to view all vertical restraints as presumptively legal. As Goldberg (1979, p. 112) has argued, "A very liberal policy toward vertical restrictions is not 'clearly best'; rather it is 'probably least worst.' " This policy prescription stems from our survey results which indicate a preponderance of either positive or not clearly negative welfare effects in combination with anticipated high costs of identifying specific cases that lead to negative welfare effects in situations that exhibit the potential for such effects.

Finally, we once again point out that public policy should provide equal treatment of practices that are economically equivalent. The differential treatment currently provided alternative vertical control mechanisms can only distort the managerial choice process in favor of safe, as opposed to efficient, strategies.

Areas for Future Research

While much has been learned over the last decade concerning the incentives for and effects of vertical integration and control, there remain, at this time, several topics of both a theoretical and an empirical nature that are underexplored. We shall briefly offer our opinions as to what these topics include.

On the theoretical side, at least five areas currently appear to require further investigation. First, the dynamic effects of vertical integration on market structure are not yet fully understood. The models of Green (1974) and Arrow (1975) indicate that such effects exist and highlight the sorts of tradeoffs between static efficiency and dynamic changes in structure that are thereby presented. It seems highly unlikely, however, that intermediate-good rationing and asymmetric information are the only sources of these dynamic effects. Other avenues remain to be explored.

Second, further theoretical analysis of the causes of partial vertical integration appear warranted. Again, several models exist that are capable of generating partial integration as an equilibrium outcome [Perry (1978c), Carlton (1979), Waterson (1982), and the disequilibrium model presented in Chapter 6], but other, perhaps more powerful, explanations of this commonly observed phenomenon are likely to exist.

Related to the topic of partial integration, a third area in need of additional theoretical work is the further analysis of the factors that influence the costs of intrafirm transfers. Throughout the literature on transaction costs (see Chapter 2) and the broader literature on organizational behavior, there are partial explanations of why internal transfer may eventually become more costly than market exchange, but a thorough treatment of this topic has yet to emerge.

A fourth area in which additional theoretical attention could prove fruitful is the analysis of what might be called mixed strategies of vertical control. In Chapter 4, we demonstrated the economic equivalence of five generic vertical control strategies under the assumptions of the variable proportions model. A mixed strategy, then, involves simultaneous use of more than one of these vertical control mechanisms. Given the demonstrated equivalence between these mechanisms, the analysis presented in Chapter 4 offers little insight concerning which strategy or combination of strategies might be preferred under a given set of market conditions. Since many of the franchise, labor union, and patent licensing agreements observed in practice make use of mixed strategies, however, an examination of the factors that determine optimal combinations of control alternatives should be of considerable interest.[4]

Finally, a fifth area that remains theoretically incomplete is the development of a model that more accurately portrays the common explanation

---

[4] Blair and Kaserman (1982b) analyzed a mixed strategy involving combined use of a lump-sum entry fee and a per-unit royalty on output.

offered by those who have pursued a vertically integrated structure—the assurance of input supplies. The models of Green (1974) and Carlton (1979) relate to this incentive but fail to capture the essence of the expressed motive. The model that perhaps comes closest to this motive is that of Klein, Crawford, and Alchian (1978). Where their model is concerned with the potential appropriation of quasi-rents by some other party through opportunistic postcontractual behavior, however, the apparent concern of many business people is a simple loss of such quasi-rents due to supply interruptions. In this latter case, the rents are not purposely appropriated by anyone: they are simply lost to all parties. Further work in this area is apparently needed.

While additional theoretical analyses of incentives for vertical integration and control remain valuable at this time, the proliferation of theories over the last decade has created a more glaring need for sound empirical work in this area. Until very recently, empirical investigations of the determinants and effects of vertical integration were severely hampered by the lack of a theoretically appealing measure of the degree of integration exhibited by a firm or an industry. A new measure recently introduced by Maddigan (1981) goes a long way toward solving this problem, although it is computationally complex. For those willing to undertake the necessary calculations, however, this approach provides a clearly superior measure of the degree to which market exchanges have been replaced by internal transfers. If should help to open this area to more meaningful empirical investigation.

Two broad types of empirical investigations are needed. The first (and clearly more difficult) sort of study involves an attempt to sort out the relative importance of the competing theories of vertical integration. With the number of potential explanations now standing at somewhere around fifteen, the need to identify which, if any, are empirically dominant is obvious to even the most casual observer.

The second broad type of empirical investigation involves the derivation and testing of hypotheses implied by the individual theories of vertical integration. That is, instead of attempting to incorporate several theories simultaneously in order to determine the relatively more important, one selects a likely theory and examines the evidence for consistency with the particular set of hypotheses implied by that theory. The recent work of Monteverde and Teece (1982a, 1982b) falls into this category. The obvious danger with this sort of endeavor is that hypotheses will be employed that are not sufficiently strong to distinguish between the alternative theories. A given empirical result may well be consistent with more than one theory of vertical integration.

Two more specific studies that would fall into this second category include (1) empirical analysis of the factors that determine the combinations of vertical control mechanisms actually used in mixed strategies (e.g., in franchise contracts), and (2) an empirical test of the potential impact of vertical integration on entry barriers. The former should rely on the further theoretical analysis of mixed strategies mentioned above. The latter would involve modeling the determinants of a new entrant's cost of capital. Clearly, much additional work remains to be done on the empirical front.

Epilogue

No story is complete without an epilogue. If literary prerogative were to allow us to conjure up our own ending, we would simply say that the convergence hypothesis was proven to be correct. The merging of opinions that we have described in Chapter 1 and attempted to verify throughout this monograph will continue until a logically consistent and evenhanded treatment of vertical integration and control emerges. A treaty will be signed by lawyers and economists that will end the hostilities in this area forever.

Such a storybook ending, however, is highly improbable. History indicates that the law is a fickle party to any treaty. Consequently, the real ending (if, indeed, one exists) will simply have to be observed by the reader.

Bibliography

Adams, W., and Dirlam, Joel B. "Steel Imports and Vertical Oligopoly Power," *American Economic Review*, Vol. 54 (September 1964), pp. 626–655.

Adelman, Morris A. "Integration and the Antitrust Laws," *Harvard Law Review*, Vol. 63 (November 1949), pp. 27–77.

Adelman, Morris A. "Concept and Statistical Measurement of Vertical Integration," in *Business Concentration and Price Policy*. National Bureau of Economic Research, Princeton, N.J.: Princeton Univ. Press, 1955.

Alchian, Armen A., and Demsetz, Harold. "Production, Information Costs, and Economic Organization," *American Economic Review*, Vol. 62 (December 1972), pp. 777–795.

Allen, Bruce T. "Vertical Integration and Market Foreclosure: The Case of Cement and Concrete," *Journal of Law and Economics*, Vol. 14 (April 1971), pp. 251–274.

Allen, R. G. D. *Mathematical Analysis for Economists*. London: Macmillan, 1938.

Areeda, Phillip, and Turner, Donald. "Predatory Pricing and Related Practices Under Section 2 of the Sherman Act," *Harvard Law Review*, Vol. 88 (February 1975), pp. 697–733.

Areeda, Phillip. "Antitrust Violations Without Damage Recoveries," *Harvard Law Review*, Vol. 89 (April 1976), pp. 1127–1139.

Areeda, Phillip, and Turner, Donald. *Antitrust Law*, Vol. III. Boston, Mass.: Little, Brown, 1978.

Areeda, Phillip, and Turner, Donald. *Antitrust Law*, Vol. IV. Boston, Mass.: Little, Brown, 1980.

Areeda, Phillip. *Antitrust Analysis*, 3d ed. Boston, Mass.: Little, Brown, 1981.

Arrow, Kenneth J. "The Organization of Economic Activity: Issues Pertinent to the Choice of Market Versus Nonmarket Allocation," in *The Analysis and Evaluation of Public Expenditures: The PPB System*, Vol. 1. Joint Economic Committee, Washington, D.C., 1969.

Arrow, Kenneth J. "Vertical Integration and Communication," *Bell Journal of Economics*, Vol. 6 (Spring 1975), pp. 173–183.

Averch, Harvey, and Johnson, Leland. "Behavior of the Firm Under Regulatory Constraint," *American Economic Review*, Vol. 52 (December 1962), pp. 1053–1069.

Bailey, Elizabeth E. and Malone, J. C. "Resource Allocation and the Regulated Firm," *Bell Journal of Economics and Management Science*, Vol. 1 (Spring 1970), pp. 129–142.

Baker, Tyler A. "The Supreme Court and the Per Se Tying Rule: Cutting the Gordian Knot," *Virginia Law Review*, Vol. 66 (November 1980), pp. 1235–1319.

Baker, Tyler A. "Interconnected Problems of Doctrine and Economics in the Section One

Labyrinth: Is *Sylvania* a Way Out?" *Virginia Law Review*, Vol. 67 (November 1981), pp. 1457–1520.

Baron, David P. "Price Uncertainty, Utility, and Industry Equilibrium in Pure Competition," *International Economic Review*, Vol. 11 (October 1970), pp. 463–480.

Barton, John H. "The Economic Basis of Damages for Breach of Contract," *Journal of Legal Studies*, Vol. 1 (June 1972), pp. 277–304.

Bauer, Joseph P. "A Simplified Approach to Tying Arrangements: A Legal and Economic Analysis," *Vanderbilt Law Review*, Vol. 33 (March 1980), pp. 283–342.

Baumol, William J., and Klevorick, Alvin K. "Input Choices and Rate-of-Return Regulation: An Overview of the Discussion," *Bell Journal of Economics and Management Science*, Vol. 1 (Autumn 1970), pp. 162–190.

Bernhardt, I. "Vertical Integration and Demand Variability," *Journal of Industrial Economics*, Vol. 25 (March 1977), pp. 213–229.

Bierman, Harold, and Tollison, Robert. "Monopoly Rent Capitalization and Antitrust Policy," *Western Economic Journal*, Vol. 8 (December 1970), pp. 385–389.

Blair, Roger D. "Random Input Prices and the Theory of the Firm," *Economic Inquiry*, Vol. 12 (June 1974), pp. 214–226.

Blair, Roger D., and Kaserman, David L. "Vertical Integration, Tying, and Antitrust Policy," *American Economic Review*, Vol. 68 (June 1978a), pp. 397–402.

Blair, Roger D., and Kaserman, David l. "Uncertainty and the Incentive for Vertical Integration," *Southern Economic Journal*, Vol. 45 (July, 1978b), pp. 266–272.

Blair, Roger D., and Kaserman, David L. "Vertical Control with Variable Proportions: Ownership Integration and Contractual Equivalents," *Southern Economic Journal*, Vol. 46 (April 1980), pp. 1118–1128.

Blair, Roger D., and Kaserman, David L. "The *Albrecht* Rule and Consumer Welfare: An Economic Analysis," *University of Florida Law Review*, Vol. 33 (Summer 1981), pp. 461–484.

Blair, Roger D., and Kaserman, David L. "Franchising: Monopoly by Contract—Comment", *Southern Economic Journal*, Vol. 48 (April 1982a) pp. 1074–1079.

Blair, Roger D., and Kaserman, David L. "A Note on Dual Input Monopoly and Tying," *Economics Letters*, Vol. 10 (1982b), pp. 145–151.

Blair, Roger D., and Kaserman, David L. "Optimal Franchising," *Southern Economic Journal*, Vol. 48 (October 1982c), pp. 494–505.

Blake, Harlan, and Jones, William. "Toward a Three-Dimensional Antitrust Policy," *Columbia Law Review*, Vol. 65 (March 1965), pp. 422–466.

Blois, K. J. "Vertical Quasi-Integration," *Journal of Industrial Economics*, Vol. 20 (1972), pp. 253–272.

Bolch, Ben W., and Damon, W. W. "The Depletion Allowance and Vertical Integration in the Petroleum Industry," *Southern Economic Journal*, Vol. 45 (July 1978), pp. 241–249.

Bork, Robert. "Vertical Integration and the Sherman Act: The Legal History of an Economic Misconception," *University of Chicago Law Review*, Vol. 22 (Autumn 1954), pp. 157–201.

Bork, Robert. "Ancillary Restraints and the Sherman Act," *A.B.A. Section of Antitrust Law*, Vol. 15 (August 1959), pp. 211–234.

Bork, Robert. "The Rule of Reason and the Per Se Concept: Price Fixing and Market Division I," *Yale Law Journal*, Vol. 74 (April 1965), pp. 775–847.

Bork, Robert. "The Rule of Reason and the Per Se Concept: Price Fixing and Market Division II," *Yale Law Journal*, Vol. 75 (January 1966a), pp. 373–475.

Bork, Robert. "Legislative Intent and the Policy of the Sherman Act," *Journal of Law and Economics,* Vol. 9 (October 1966b), pp. 7–48.

Bork, Robert. "Vertical Integration and Competitive Processes," in *Public Policy Toward Mergers,* edited by J. F. Weston and S. Peltzman. Pacific Palisades, California: Goodyear Publishing Company, 1969.

Bork, Robert. "Vertical Restraints: *Schwinn* Overruled," in *1977 Supreme Court Review,* edited by Philip Kurland. Chicago, Ill.: University of Chicago Press, 1978a.

Bork, Robert. *The Antitrust Paradox.* New York: Basic Books, Inc., 1978b.

Bowley, A. L. "Bilateral Monopoly," *Economic Journal,* Vol. 28 (December 1928), pp. 651–659.

Bowman, Ward. "Prerequisites and Effects of Resale Price Maintenance," *University of Chicago Law Review,* Vol. 22 (Summer 1955), pp. 825–873.

Bowman, Ward. "Tying Arrangements and the Leverage Problem," *Yale Law Journal,* Vol. 67 (November 1957), pp. 10–36.

Burstein, Meyer. "A Theory of Full-Line Forcing," *Northwestern University Law Review,* Vol. 55 (March/April 1960), pp. 62–95.

Canes, Michael E. "The Vertical Integration of Oil Firms," in *Resource Allocation and Economic Policy,* edited by M. L. Burstein and M. Allingham. London: Macmillan, 1976.

Carlton, Dennis W. "Vertical Integration in Competitive Markets Under Uncertainty," *Journal of Industrial Economics,* Vol. 27 (March 1979), pp. 189–209.

Caves, Richard E., and Murphy, II, William F. "Franchising: Firms, Markets, and Intangible Assets," *Southern Economic Journal,* Vol. 42 (April 1976), pp. 572–586.

Coase, Ronald. "The Nature of the Firm," *Economica,* Vol. 4 (November 1937), pp. 386–405; reprinted in *Readings in Price Theory,* edited by George J. Stigler and Kenneth Boulding. Homewood, Ill.: Richard D. Irwin, Inc. 1952, pp. 331–351.

Comanor, William S. "Vertical Mergers, Market Power, and the Antitrust Laws," *American Economic Review,* Vol. 57 (May 1967), pp. 254–265.

Comanor, William S. "Vertical Territorial and Customer Restrictions: *White Motor* and Its Aftermath," *Harvard Law Review,* Vol. 81 (May 1968), pp. 1419–1438.

Crandall, Robert. "Vertical Integration and the Market for Repair Parts in the United States Automobile Industry," *Journal of Industrial Economics,* Vol. 16 (July 1968), pp. 212–234.

Cummings, F. Jay, and Ruhter, Wayne E. "The *Northern Pacific* Case," *Journal of Law and Economics,* Vol. 22 (October 1979), pp. 329–350.

Dayan, David, "Behavior of the Vertically Integrated Firm Under Regulatory Constraint," Paper presented at the Econometric Society Winter Meeting, New York, 1973.

Dayan, David, "Behavior of the Firm Under Regulatory Constraint: A Reexamination," *Industrial Organization Review,* Vol. 3 (1975), pp. 61–76.

Dirlam, Joel, and Kahn, Alfred E. *Fair Competition.* Ithaca, N.Y.: Cornell University Press, 1954.

Dixit, Avinash, and Norman, Victor. "Advertising and Welfare," *Bell Journal of Economics,* Vol. 9 (Spring 1978), pp. 1–17.

Easterbrook, Frank H. "Maximum Price Fixing," *University of Chicago Law Review,* Vol. 48, (Fall 1981), pp. 886–910.

Edwards, Corwin D. "Vertical Integration and the Monopoly Problem," *Journal of Marketing,* Vol. 17 (April 1953), pp. 404–410.

Fama, Eugene F. "Agency Problems and the Theory of the Firm," *Journal of Political Economy,* Vol. 88 (April 1980), pp. 288–307.

Fellner, William. "Prices and Wages Under Bilateral Monopoly," *Quarterly Journal of Economics*, Vol. 61 (August 1947), pp. 503–509.

Ferguson, C. E. *The Neoclassical Theory of Production and Distribution*. Cambridge, England: Cambridge Univ. Press, 1969.

Friedman, Milton, and Savage, Leonard J. "The Utility Analysis of Choices Involving Risk," *Journal of Political Economy*, Vol. 56 (August 1948), pp. 279–304.

Goldberg, Victor P. "The Law and Economics of Vertical Restrictions: A Relational Perspective," *Texas Law Review*, Vol. 58 (December 1979), pp. 91–129.

Gould, J. R. "Price Discrimination and Vertical Control: A Note," *Journal of Political Economy*, Vol. 85 (October 1977), pp. 1063–1071.

Green, Jerry R. "Vertical Integration and Assurance of Markets," Discussion Paper Number 383, Harvard Institute of Economic Research, October 1974.

Greenhut, M. L., and Ohta, H. "Related Market Conditions and Interindustrial Mergers," *American Economic Review*, Vol. 66 (June 1976), pp. 267–277.

Greenhut, M. L., and Ohta, H. "Related Market Conditions and Interindustrial Mergers: Reply," *American Economic Review*, Vol. 68 (March 1978), pp. 228–230.

Greenhut, M. L., and Ohta, H. "Vertical Integration of Successive Oligopolists," *American Economic Review*, Vol. 69 (March 1979), pp. 137–141.

Handler, Milton. "Changing Trends in Antitrust Doctrines: An Unprecedented Supreme Court Term—1977," *Columbia Law Review*, Vol. 77 (November 1977), pp. 979–1028.

Hansen, Robert S., and Roberts, R. Blaine. "Metered Tying Arrangements, Allocative Efficiency, and Price Discrimination," *Southern Economic Journal*, Vol. 47 (July 1980), pp. 73–83.

Haring, John R., and Kaserman, David L. "Related Market Conditions and Interindustrial Mergers: Comment," *American Economic Review*, Vol. 68 (March 1978), pp. 225–227.

Hay, George. "An Economic Analysis of Vertical Integration," *Industrial Organization Review*, Vol. 1 (1973), pp. 188–198.

Horowitz, Ira. *Decision Making and the Theory of the Firm*. New York: Holt, 1970.

Inaba, Frederick S. "Franchising: Monopoly by Contract," *Southern Economic Journal*, Vol. 47 (July 1980), pp. 65–72.

Irwin, Manley R., and McKee, R. E. "Vertical Integration and the Communication Equipment Industry: Alternatives for Public Policy," *Cornell Law Review*, Vol. 52 (February 1968), pp. 446–472.

Jensen, H. R., Kehrberg, E. W., and Thomas, D. W. "Integration as an Adjustment to Risk and Uncertainty," *Southern Economic Journal*, Vol. 28 (1962), pp. 378–384.

Jones, William K. "The Two Faces of *Fortner:* Comment on a Recent Antitrust Opinion," *Columbia Law Review*, Vol. 78 (January 1978), pp. 39–47.

Just, Richard E., and Salkin, M. S. "Welfare Effects of Stabilization in a Vertical Market Chain," *Southern Economic Journal*, Vol. 42 (April 1976), pp. 633–643.

Kahn, Alfred E. *The Economics of Regulation*, Vol. II. New York: Wiley, 1971.

Kaserman, David L. "Theories of Vertical Integration: Implications for Antitrust Policy," *Antitrust Bulletin*, Vol. 23 (Fall 1978), pp. 483–510.

Kessler, F., and Stern, R. H. "Competition, Contract, and Vertical Integration," *Yale Law Journal*, Vol. 69 (November 1959), pp. 1–129.

Kitch, Edmund W. "The Yellow Cab Antitrust Case," *Journal of Law and Economics*, Vol. 15 (October 1972), pp. 327–336.

Klein, Benjamin, Crawford, Robert G., and Alchian, Armen A. "Vertical Integration, Appropriable Rents, and the Competitive Contracting Process," *Journal of Law and Economics*, Vol. 21 (October 1978), pp. 297–326.

Klein, Benjamin. "Transaction Cost Determinants of 'Unfair' Contractual Arrangements," *American Economic Review*, Vol. 70 (May 1980), pp. 356–362.

Laffer, Arthur B. "Vertical Integration by Corporations, 1929–1965," *Review of Economics and Statistics*, Vol. 51 (February 1969), pp. 91–93.

Letwin, William. "Congress and the Sherman Antitrust Law: 1877–1890," *University of Chicago Law Review*, Vol. 23 (Winter 1956), pp. 221–258.

Markovits, Richard S. "Tie-Ins and Reciprocity: A Functional, Legal, and Policy Analysis," *Texas Law Review*, Vol. 58 (November 1980), pp. 1363–1445.

Marshall, Alfred. *Principles of Economics*, 8th ed. London: Macmillan, 1930.

Marvel, Howard P. "Exclusive Dealing," *Journal of Law and Economics*, Vol. 25 (April 1982), pp. 1–26.

McGee, John S., and Bassett, Lowell R. "Vertical Integration Revisited," *Journal of Law and Economics*, Vol. 19 (April 1976), pp. 17–38.

McGee, John S. "Predatory Pricing Revisited," *Journal of Law and Economics*, Vol. 23 (October 1980), pp. 289–330.

McKenzie, Lionel W. "Ideal Output and the Interdependence of Firms," *Economic Journal*, Vol. 61 (December 1951), pp. 785–803.

Meehan, James W., and Larner, Robert J. "A Proposed Rule of Reason for Vertical Restraints on Competition," *Antitrust Bulletin*, Vol. 26 (Summer 1981), pp. 195–225.

Mitchell, E. J., "Capital Cost Savings of Vertical Integration," *Vertical Integration in the Oil Industry*, edited by E. J. Mitchell. Washington, D.C.: American Enterprise Institute, 1976.

Monteverde, Kirk, and Teece, David J. "Supplier Switching Costs and Vertical Integration in the Automobile Industry," *Bell Journal of Economics*, Vol. 13 (Spring 1982), pp. 206–213.

Monteverde, Kirk, and Teece, David J. "Appropriable Rents and Quasi-Vertical Integration," *Journal of Law and Economics*, Vol. 25 (October 1982), pp. 321–328.

Mueller, Willard F. "Public Policy Toward Vertical Mergers," *Public Policy Toward Mergers*, edited by J. F. Weston and S. Peltzman. Pacific Palisades Calif.: Goodyear Publishing Company, 1969.

Neale, A. D., and Goyder, D. G. *The Antitrust Laws of the U.S.A.*, 3rd ed. New York: Cambridge University Press, 1980.

Note, "Newcomer Defenses: Reasonable Use of Tie-ins, Franchises, Territorials, and Exclusives," *Stanford Law Review*, Vol. 18 (January 1966), pp. 457–474.

Note, "Total-Sales Royalties Under the Patent-Misuse Doctrine: A Critique of *Zenith*," *Michigan Law Review*, Vol. 76 (June 1978), pp. 1144–1176.

Pashigian, B. Peter. *The Distribution of Automobiles, An Economic Analysis of the Franchise System*. Prentice-Hall: Englewood Cliffs, N.J. 1961.

Patinkin, Don. "Multiple-Plant Firms, Cartels, and Imperfect Competition," *Quarterly Journal of Economics*, Vol. 61 (February 1947), pp. 173–205.

Peltzman, Sam. "Issues in Vertical Integration Policy," in *Public Policy Toward Mergers*, edited by J. F. Weston and S. Peltzman. Pacific Palisades, Calif.: Goodyear Publishing Company, 1969.

Perry, Martin K. "The Theory of Vertical Integration by Imperfectly Competitive Firms," Center for Research in Economic Growth Memorandum Number 197, Stanford University, 1975.

Perry, Martin K. "Price Discrimination and Forward Integration," *Bell Journal of Economics*, Vol. 9 (Spring 1978a), pp. 209–217.

Perry, Martin K. "Related Market Conditions and Interindustrial Mergers: Comment," *American Economic Review*, Vol. 68 (March 1978b), pp. 221–224.

Perry, Martin K. "Vertical Integration: The Monopsony Case," *American Economic Review*, Vol. 68 (Sept. 1978c), pp. 561–570.

Perry, Martin K. "Vertical Integration by Competitive Firms: Uncertainty and Diversification," *Southern Economic Journal*, Vol. 49 (July 1982), pp. 201–208.

Peterman, John L. "The *International Salt* Case," *Journal of Law & Economics*, Vol. 22 (October 1979), pp. 351–364.

Pitofsky, Robert. "The *Sylvania* Case: Antitrust Analysis of Non-Price Vertical Restrictions," *Columbia Law Review*, Vol. 78 (January 1978), pp. 1–38.

Posner, Richard A. "Exclusionary Practices and the Antitrust Laws," *University of Chicago Law Review*, Vol. 41 (Spring 1974), pp. 506–535.

Posner, Richard A. *Antitrust Law: An Economic Perspective.* Chicago, Ill.: University of Chicago Press, 1976.

Posner, Richard A. "The Rule of Reason and the Economic Approach: Reflections on the *Sylvania* Decision," *University of Chicago Law Review*, Vol. 45 (Fall 1977), pp. 1–20.

Posner, Richard A. "The Next Step in the Antitrust Treatment of Restricted Distribution: Per Se Legality," *University of Chicago Law Review*, Vol. 48 (Winter 1981), pp. 6–26.

Qualls, P. David. "Market Structure Theory and the Policy Implications of Product Complementarity," *Industrial Organization Review*, Vol. 6 (1978), pp. 38–48.

Slawson, W. David. "A Stronger, Simpler Tie-In Doctrine," *Antitrust Bulletin*, Vol. 25 (Winter 1980), pp. 671–699.

Smith, Vernon L. "The Borrower-Lender Contract Under Uncertainty," *Western Economic Journal*, Vol. 9 (March 1971), pp. 52–56.

Spence, A. Michael. "Entry, Capacity, Investment and Oligopolistic Pricing," *Bell Journal of Economics*, Vol. 8 (Autumn 1977), pp. 534–544.

Spengler, Joseph J. "Vertical Integration and Antitrust Policy," *Journal of Political Economy*, Vol. 58 (July/August 1950), pp. 347–352.

Stigler, George J. "The Division of Labor is Limited by the Extent of the Market," *Journal of Political Economy*, Vol. 59 (June 1951), pp. 185–193.

Stigler, George J. "A Note on Block Booking," edited by Philip Kurland. *The Supreme Court Review*, 1963.

Stigler, George. "Price and Non-Price Competition," *Journal of Political Economy*, Vol. 76 (February 1968), pp. 149–154.

Strasser, Kurt A. "Vertical Territorial Restraints After *Sylvania*: A Policy Analysis and Proposed New Rule," *Duke Law Journal*, Vol. 1977 (October 1977), pp. 775–840.

Schildkraut, Marc G. "Areas of Primary Responsibility and Other Territorial Restrictions in Channels of Distribution Under Antitrust Laws: A Legal and Economic Analysis," *Columbia Journal of Law and Social Problems*, Vol. 11 (Summer 1975), pp. 509–570.

Schmalensee, Richard. "A Note on the Theory of Vertical Integration," *Journal of Political Economy*, Vol. 81 (March/April 1973), pp. 442–449.

Schupack, Mark B. "The Theory of Vertical Integration: A Survey," Working Paper No. 77-31, Department of Economics, Brown University, October 1977.

Sheahan, John B. "Integration and Exclusion in the Telephone Equipment Industry," *Quarterly Journal of Economics*, Vol. 70 (May 1956), pp. 249–269.

Shelton, John P. "Allocative Efficiency v. 'X'-Efficiency: Comment," *American Economic Review*, Vol. 57 (December 1967), pp. 1252–1258.

Shillinglaw, Gordon. "The Effects of Requirements Contracts on Competition," *Journal of Industrial Economics*, Vol. 2 (April 1954), pp. 147–163.

Singer, Eugene M. *Antitrust Economics.* Englewood Cliffs, N.J.: Prentice-Hall, 1968.

Singer, Eugene M. *Antitrust Economics and Legal Analysis.* Columbus, Ohio: Grid Publishing, Inc., 1981.

Sullivan, Lawrence A. *Handbook of the Law of Antitrust*. St. Paul, Minn.: West Publishing Co., 1977.

Takayama, Akira. "Behavior of the Firm Under Regulatory Constraint," *American Economic Review*, Vol. 59 (June 1969), pp. 255–260.

Telser, Lester G. "Why Should Manufacturers Want Fair Trade?" *Journal of Law and Economics*, Vol. 3 (October 1960), pp. 86–105.

Telser, Lester G. "A Theory of Monopoly of Complementary Goods," *Journal of Business*, Vol. 52 (April 1979), pp. 211–230.

Thorelli, Hans B. *The Federal Antitrust Policy*. London: Allen & Unwin, 1954.

Tucker, Irwin B., and Wilder, Ronald P. "Trends in Vertical Integration in the U.S. Manufacturing Sector," *Journal of Industrial Economics*, Vol. 26 (September 1977), pp. 81–94.

Turner, Donald. "Conglomerate Mergers and Section 7 of the Clayton Act," *Harvard Law Review*, Vol. 78 (May 1965), pp. 1313–1395.

Vernon, John, and Graham, Daniel. "Profitability of Monopolization by Vertical Integration," *Journal of Political Economy*, Vol. 79 (July/August 1971), pp. 924–925.

Wallace, D. H. *Market Control in the Aluminum Industry*. Cambridge, Mass.: Harvard Univ. Press, 1937.

Warren-Boulton, Frederick R. "Vertical Control with Variable Proportions," *Journal of Political Economy*, Vol. 82 (July/August 1974), pp. 783–802.

Warren-Boulton, Frederick R. "Vertical Control by Labor Unions," *American Economic Review*, Vol. 67 (June 1977), pp. 309–322.

Warren-Boulton, Frederick R. *Vertical Control of Markets: Business and Labor Practices*. Cambridge, Mass.: Ballinger Publishing Co., 1978.

Waterson, Michael. "Price-Cost Margins and Successive Market Power," *Quarterly Journal of Economics*, Vol. 94 (1980), pp. 135–150.

Waterson, Michael. "Vertical Integration, Variable Proportions and Oligopoly," *Economic Journal*, Vol. 92 (March 1982), pp. 129–144.

Westfield, Fred M. "Vertical Integration: Does Product Price Rise or Fall?" *American Economic Review*, Vol. 71 (June 1981), pp. 334–346.

White, Lawrence J. "Vertical Restraints in Antitrust Law: A Coherent Model," *Antitrust Bulletin*, Vol. 26 (Summer 1981), pp. 327–345.

Williamson, Oliver E. "Economics As An Antitrust Defense: The Welfare Tradeoffs," *American Economic Review*, Vol. 58 (March 1968), pp. 18–34.

Williamson, Oliver E. "The Vertical Integration of Production: Market Failure Considerations," *American Economic Review*, Vol. 61 (May 1971), pp. 112–123.

Williamson, Oliver E. "Markets and Hierarchies: Some Elementary Considerations," *American Economic Review*, Vol. 63 (May 1973), pp. 316–325.

Williamson, Oliver E. "The Economics of Antitrust: Transactions Cost Considerations," *University of Pennsylvania Law Review*, Vol. 122 (May 1974), pp. 1429–1496.

Williamson, Oliver E. *Markets and Hierarchies: Analysis and Antitrust Implications*. New York: The Free Press, 1975.

Williamson, Oliver E. "Assessing Vertical Market Restrictions: Antitrust Ramifications of the Transaction Cost Approach," *University of Pennsylvania Law Review*, Vol. 127 (April 1979), pp. 953–993.

Wu, S. Y. "The Effects of Vertical Integration on Price and Output," *Western Economic Journal*, Vol. 2 (1964), pp. 117–133.

Zajac, Edward E. "A Geometric Treatment of Averch-Johnson's Behavior of the Firm Model," *American Economic Review*, Vol. 60 (March 1970), pp. 117–125.

Zelek, E. F., Stern, L. W., and Dunfee, T. W. "A Rule of Reason Decision Model After Sylvania," *California Law Review*, Vol. 68 (January 1980), pp. 13–47.

Index